Basic Statistics

James Snowdon
B.A., F.C.C.S., F.F.T.Com.

Head of Department of Commerce & Business Studies
College of Further Education, Kirkby, Lancashire

Sometime Lecturer in Statistics:
Stretford Technical College, Lancs.
Salford Technical College, Lancs.

LONDON
BUTTERWORTHS
1969

ENGLAND: BUTTERWORTH & CO (PUBLISHERS) LTD.
LONDON: 88 KINGSWAY, W.C.2.

AUSTRALIA: BUTTERWORTH & CO. (AUSTRALIA) LTD.
SYDNEY: 20 LOFTUS STREET
MELBOURNE: 343 LITTLE COLLINS STREET
BRISBANE: 240 QUEEN STREET

CANADA: BUTTERWORTH & CO. (CANADA) LTD.
TORONTO: 14 CURITY AVENUE, 16

NEW ZEALAND: BUTTERWORTH & CO. (NEW ZEALAND) LTD.
WELLINGTON: 49/51 BALLANCE STREET
AUCKLAND: 35 HIGH STREET

SOUTH AFRICA: BUTTERWORTH & CO. (SOUTH AFRICA) LTD.
DURBAN: 33/35 BEACH GROVE

Standard Book Number: 406 78650 X

Printed in Great Britain by Northumberland Press Limited, Gateshead

To my wife—who has always wondered what I teach!

Preface

For some years there were very few books in this subject which were really suitable for technical colleges and students had to use university or final-professional texts. I have felt for some time that a book was needed for the student who hates mathematics and who views the prospect of having to learn 'Statistics' with even greater horror. This volume is specifically aimed at such a market.

To make the subject palatable to such readers I have kept the language of the text as simple as possible and have avoided formulas with Greek letters as far as possible. I have kept to the O.N.C./O.N.D. Guide Syllabus as outlined in Rules 124 (Revised 1967) and resisted the temptation to cover 'O' level and other syllabuses of similar content.

This will not prevent colleagues who feel their students can take rather more from adding further explanatory notes as they go along and extending the text at the end to cover G.C.E. and similar examinations.

The methods outlined in this volume have been tried and tested for a decade or so by the author and I hope the work will help to foster an interest in the subject among the students it is intended to help.

I acknowledge with grateful thanks the permission of the Union of Lancashire and Cheshire Institutes and the Associated Examining Board to quote examples from past examination papers in the exercises. I also wish to thank my friend and colleague Mr. J. E. Hillidge B.A. for checking the manuscript.

J.S.

Contents

Introduction

WHY STUDY STATISTICS?

Why do we study this subject? The simple answer is that it helps us to make the best decisions in business.

If you found that your firm produced and sold 500 components in a certain week, this fact would mean little on its own. If you then discovered that 600 were normally produced in a week, you would want to know why the production and sales had fallen. The reason might be quite simple and be acceptable to you but you would find out what the reason was. Likewise the discovery that normal production was only 400 would give you some pleasure but it would not be wise to rest on your laurels. You would find out why the increase had come about and see if such improvement could be maintained. A wise manager would further see how the 500 figure compared with other firms of similar size in the industry. This is the kind of study which a knowledge of Statistics helps. You will discover what figures to look at and how to appraise them. You will learn how to see any obvious errors or flaws in them which might otherwise cloud sane judgment and you will learn quick methods for calculating the measures we use in the subject.

Statistics is also useful in forward planning. A clear upward trend in production and sales might make it clear that an extension to the factory would have to be built before it was too late to do so without serious disruption of everyday work. In this way the statistician can aid the accountant and the directors by saving money and resources for the firm.

Statistics is nevertheless not an exact science. It is a science of approximation—the science of the 'best guess'. It deals only with those aspects of a problem which can be measured and will thus never give the whole answer. It may often give no definite correct answer but it will provide the basis for making a judgment which is as sound as possible in the given circumstances.

When you studied mathematics your textbooks would always agree on a single correct answer to a problem even if the methods of solution varied. In statistics you must not be surprised if two textbooks with varying methods of dealing with a problem subsequently produce two answers to it. You may likewise find that different lecturers hold separate opinions on a given topic. In such cases remember the limitations of the subject and try to understand that your job as a statistician will be to produce your 'best guess' and, in cutting through the jungle of complexities, to explain that opinion in layman's terms to your colleagues and get your postulation accepted because it is understood and it seems the right one.

1

Frequency Distributions

If we took the ages of one hundred men and boys at random from a crowd we might find the results would be as follows:

3	47	43	73	86	36	96	47	36	61
97	74	24	67	62	42	81	14	57	20
16	76	62	27	66	56	50	26	71	7
12	56	85	99	26	96	96	68	27	31
55	59	56	35	64	38	54	82	46	22
16	22	77	94	39	49	54	43	54	82
84	42	17	53	31	57	14	55	6	88
63	9	63	78	59	16	95	55	67	19
33	21	12	34	29	78	64	56	8	82
57	60	86	32	64	9	47	27	96	54

It is very difficult to see at a glance just what sort of range or pattern the ages we have when we see the figures in this form.

To make it easier we decide on a number of groups and then add up how many ages fall in each group. Let us suppose we decide to call the first group 1 to 10, the second 11 to 20 and so on. The number of different possible readings in each group is called the group interval so in this case the group interval is ten. Ten such groups would cover all the ages because we have a range of a hundred years and we start a new group every ten years.

We must now set out our ten groups and put a vertical mark by each group every time an age falls within its span. For convenience in the eventual addition we cross through each four lines when we have our fifth reading instead of adding a fifth stroke. When we have finished this task we total the frequencies.

This table gives us a better idea of the pattern of ages in the frequency distribution. It appears that there are a few boys, a steady number of men up to 50, a lot of men between 51 and 70, and then a tailing off.

We can illustrate these figures by setting out the groups on the horizontal axis of a piece of graph paper and the frequencies on the vertical axis, *see* Figure 1. Do remember to label the axes and title the diagram.

Ages of a hundred men and boys

Groups	*Frequency*	*Total*
0-10	~~IIII~~ I	6
11-20	~~IIII~~ ~~IIII~~	10
21-30	~~IIII~~ ~~IIII~~	10
31-40	~~IIII~~ ~~IIII~~	10
41-50	~~IIII~~ ~~IIII~~	10
51-60	~~IIII~~ ~~IIII~~ ~~IIII~~ III	18
61-70	~~IIII~~ ~~IIII~~ II	12
71-80	~~IIII~~ II	7
81-90	~~IIII~~ IIII	9
91-100	~~IIII~~ III	8

This diagram is known as a block diagram or histogram. By joining up the mid-points of the centres of the blocks we can represent the information by means of a rough line known as a frequency polygon but you should not do this unless the examination question requires it.

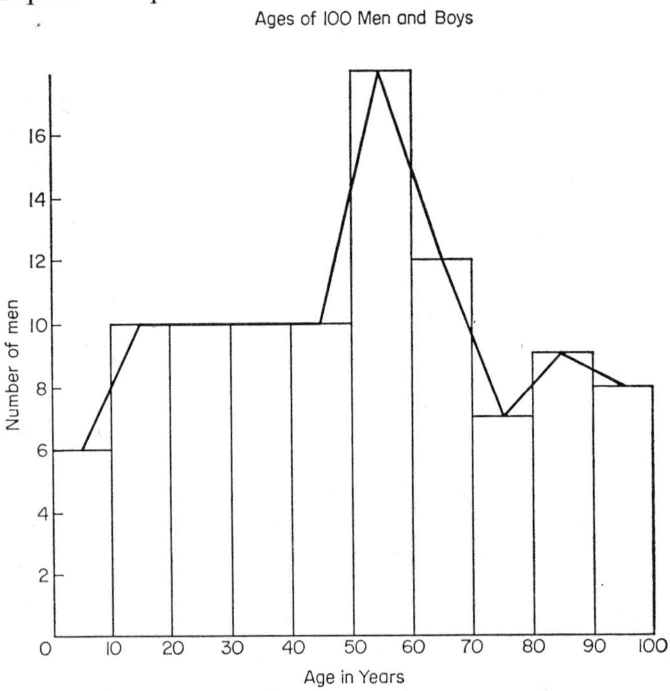

Fig. 1.—Block diagram and frequency polygon

2

Cumulative Frequency Distributions

In a frequency distribution it is sometimes more useful to know at any given point in the distribution how many readings we have had so far. For instance it might be useful to know how many of the men were under 50 or how many were under 60. In such cases we cumulate the total frequencies by totalling at the end of each line all the readings we have so far; thus:

Groups	*Frequency*	*Cumulative Frequency*	
0-10	6		6
11-20	10	(6+10=)	16
21-30	10	(6+10+10=)	26
31-40	10	(6+10+10+10)	36
41-50	10	etc.	46
51-60	18		64
61-70	12		76
71-80	7		83
81-90	9		92
91-100	8		100

It will be noted that the last figure in the cumulative frequency column is a hundred which is the same as the total of men in the distribution. This is a good way of checking that your mental arithmetic in the said column is accurate.

Just as we normally produce a block diagram or histogram from a plain distribution, so we produce a cumulative frequency drive or ogive from a cumulative frequency distribution.

Again we put the groups on the horizontal axis and the frequencies on the vertical axis and plot our curve as illustrated in Figure 2.

Fig. 2.—Cumulative frequency curve

3

The Mode

In Statistics there are a number of different averages which we use on appropriate occasions. The Mode is one of these. It is *the item which occurs most frequently in the distribution* and is thus the best way of giving the 'typical item'. It is the kind of average meant in such phrases as 'the average man takes a two week holiday every year'. It means that the usual holiday is two weeks. If the age of thirteen boys in a crowd were:

7, 11, 10, 14, 11, 9, 10, 8, 12, 11, 13, 11, 14

then the Mode would be 11 for it is the commonest reading in the distribution. It appears four times whilst the others appear once or twice only.

There is an arithmetical method of finding the Mode but this is not normally required. The usual way is the drawing method as follows:

(1) Draw a block diagram from the distribution.

(2) Draw a line from the top left hand corner of the tallest block to the top left hand corner of the next block on the right.

(3) Draw another line from the top right hand corner of the tallest block to the top right hand corner of the next block on the left.

(4) From the point where the two lines intersect, drop a perpendicular on to the horizontal axis. At the spot where the perpendicular crosses the axis you can read off the Mode. Label the diagram fully and mark in the Mode clearly.

In Figure 3 the ages of the thirteen boys are illustrated by a block diagram and the method above has been used to find the Mode. Follow the above instructions point by point on the diagram and make sure that you understand what has been done. You will notice that the answer includes a decimal and is, in fact, a figure for which there is no definite reading in the distribution at all! In Statistics such an answer is acceptable for we are not dealing with an exact science but a science of approximation.

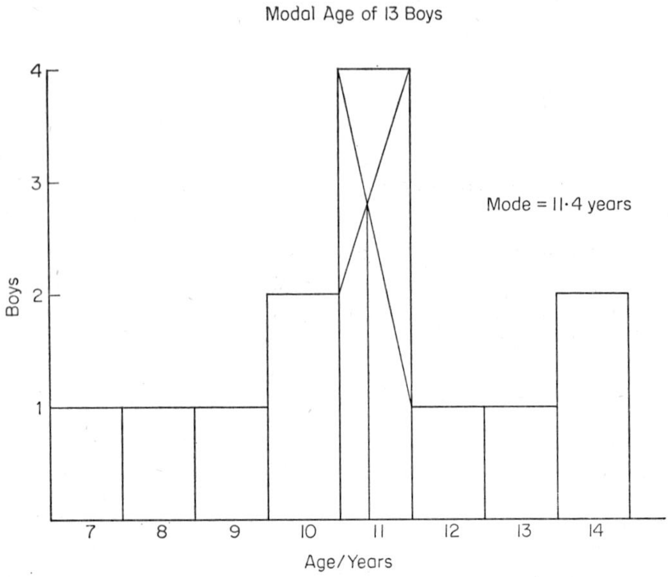

Fig. 3.—Block diagram used to find Mode

4

The Median

The Median is the second of the three important statistical averages which we must know. It is the *middle item of the distribution when all the items are arranged in order*. If the thirteen boys mentioned in the previous chapter were made to stand in order of height, the Median height would be that of the seventh boy, the one in the middle, who would have equal numbers of boys on either side of him. The Median is not only useful when we wish to know the value of the middle reading in the distribution. Its calculation is not influenced by extreme values at either end of the scale so it is useful where such extreme values might make other assessment difficult. To give an extreme example let us suppose that the ages of five men were 22, 23, 27, 21 and 84. It is clear that the 84-year-old would upset calculations otherwise based on very young men so the Median would be appropriate here.

There are two methods of finding the Median of a distribution and both must be known.

(1) THE DRAWING METHOD

(*a*) Draw a Cumulative Frequency Curve or Ogive from the data given by the method outlined in Chapter 2.

(*b*) Decide where the Median will lie in the distribution by using the formula $\frac{n+1}{2}$ where 'n' is the number of items in the distribution.

(*c*) Mark the position thus calculated on the vertical axis of the Ogive and produce a horizontal from that point until it crosses the Ogive.

(*d*) Drop a perpendicular from the point of intersection on to the horizontal axis and read off the Median at that point.

EXAMPLE Let us suppose that the heights of thirteen men, chosen at random from a crowd, are as follows:

Height of Man to the Nearest Inch	*Number of Men*
5 ft. 6 in.	1
5 ft. 7 in.	1
5 ft. 8 in.	1
5 ft. 9 in.	2
5 ft. 10 in.	3
5 ft. 11 in.	2
6 ft. 0 in.	2
6 ft. 1 in. and over	1

It is clear on inspection that the Median is 5 ft. 10 in. for the seventh man in height order is clearly one of the three men with a height of 5 ft. 10 in. In many distributions however the fact is not so obvious so let us apply the drawing method as outlined and calculate the Median from our Ogive.

(*a*) In Figure 4 you will see that an Ogive has been drawn from the table giving the heights of our thirteen men.

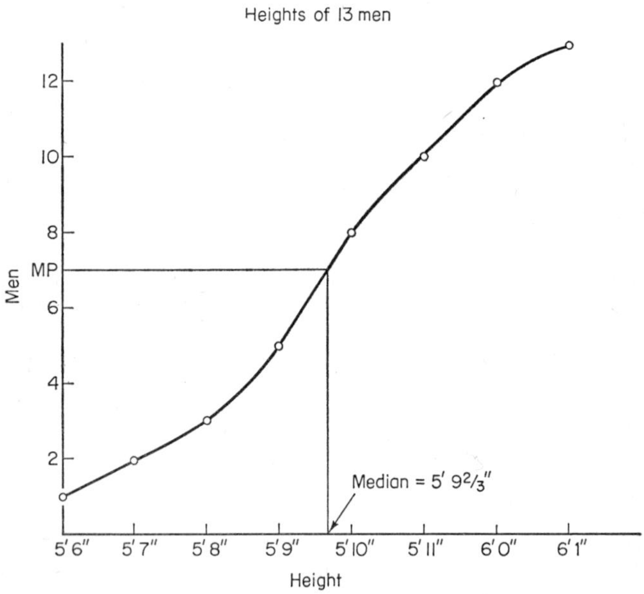

Fig. 4.—Ogive used to find Median

(*b*) Using the formula $\frac{n+1}{2}$ where n is the number of items in the distribution we find that the Median will lie at the $\frac{13+1}{2}=7$th position so:

(*c*) We mark the seventh position on the vertical axis and produce a horizontal until it crosses the Ogive.

(*d*) We then drop a perpendicular from the point of intersection on to the horizontal axis and read off the Median at that point.

We must now consider the arithmetic method. Examination questions usually make it clear which method is required.

(2) THE ARITHMETIC METHOD

(*a*) Set out the Cumulative Frequency Table, find the position of the Median using $\frac{n+1}{2}$ as before and mark it with a small arrow in the Cumulative Frequency Column.

(*b*) Calculate the Median using the following formula:

$$\text{GREATEST POSSIBLE READING IN THE GROUP BEFORE THE ARROW} + \frac{\text{MEDIAN POSITION MINUS CUMULATIVE FREQUENCY OF GROUP BEFORE THE ARROW}}{\text{ORDINARY FREQUENCY OF THE GROUP AFTER THE ARROW}} \times \text{GROUP INTERVAL}$$

EXAMPLE We can best differentiate between the two methods by using the same figures of the heights of men for the arithmetic method.

(*a*) *Height of Man to the Nearest Inch Below*	*No. of Men*	*Cumulated*
5 ft. 6 in.	1	1
5 ft. 7 in.	1	2
5 ft. 8 in.	1	3
5 ft. 9 in.	2	5
		←MP=7
5 ft. 10 in.	3	8
5 ft. 11 in.	2	10
6 ft. 0 in.	2	12
6 ft. 1 in. and over	1	13

(*b*) The group before the arrow is the one marked '5 ft. 9 in.' to the nearest inch below and the highest possible reading in this group would be a man so very slightly less than 5 ft. 10 in. tall that, to all intents and purposes, we must call him 5 ft. 10 in. This then becomes the highest possible reading in the group before the arrow. The Median position is 7 and the Cumulative Frequency of the group before the arrow is 5 so subtracting them gives us 2. The ordinary frequency of the group after the arrow is 3. The Group Interval refers to how often a new group is started. The answer in this case is one inch. The formula thus gives us:

$$5 \text{ ft. } 9 \text{ in.} + \tfrac{2}{3} \times 1 \text{ in.}$$

which gives the Median as five feet, nine and two thirds inches.

5

Quartiles, Deciles and Percentiles

Just as the Median tells us of the value of the middle reading in the distribution so the lower quartile or first quartile (Q1) tells us the value of the reading one quarter of the way through the distribution and the upper quartile or third quartile (Q3) tells us of the value of the reading three quarters of the way through. The position of the lower quartile is found from the formula $\frac{n+1}{4}$ and that of the third quartile from the formula $\frac{3(n+1)}{4}$. The measures can then be found either by the drawing method or the arithmetic method by following the rules given in the previous chapter.

Similarly the readings at tenths of the way through the distribution are called the first decile, second decile and so on up to the ninth decile though we do not speak of the fifth decile for that is the median. The formula for finding the position of the deciles is $\frac{?n+1}{10}$ where ? is the number of the decile required. The eighth decile would thus be found $\frac{8(n+1)}{10}$. The readings at hundredths of the way through the distribution are called first percentile, second percentile etc. up to the ninety-ninth percentile though, once again, we do not usually speak of the fiftieth percentile for it is the same as the Median and the twenty-fifth and seventy-fifth percentiles are the same as the Quartiles. Likewise every even tenth percentile will usually keep its decile name. The formula for finding the position of the percentiles is $\frac{?n+1}{100}$ where ? is the number of the percentile required. Thus the position of the forty-third percentile would be found from $\frac{43(n+1)}{100}$.

EXAMPLE Let us suppose that the table below records the marks which a hundred students received in their Statistics examination.

It was decided to fail 35 per cent of the students and give distinctions to the top 10 per cent. What mark is the level for pass and for distinction? The first thing we do is to cumulate the frequency. Then we decide which measures we need. To fail 35 per cent we we shall need to find the thirty-fifth percentile. To give distinction to the top ten per cent we need the mark ten per cent from the top so we shall want the ninth decile. In examination questions it is usually quite clear whether the drawing method or arithmetic method is required. We will do it both ways for practice beginning with the arithmetic method.

The position of the thirty-fifth percentile will be at $\frac{35(101)}{100}$ and the ninth decile will appear at $\frac{9(101)}{10}$. Answers are 35·35 and 90·9

Percentage Mark	*Number of Students* f	*Cumulative Frequency* cf
0-10	4	4
11-20	7	11
21-30	11	22
		←P35P=35·35
31-40	15	37
41-50	21	58
51-60	19	77
61-70	10	87
		←P9D=90·9
71-80	7	94
81-90	4	98
91-100	2	100

Using the formula given in the Median chapter we find:

$$P35 = 30 + \frac{13{\cdot}35}{15} \times 10 = 38{\cdot}9$$

and

$$D9 = 70 + \frac{3{\cdot}9}{7} \times 10 = 75{\cdot}57$$

Hence students would pass at 38·9 per cent (probably meaning 39 per cent would be taken) and they would gain Distinction at 75·57 per cent (probably meaning 76 per cent). You can compare these answers with the answers obtained by the drawing method in Figure 5.

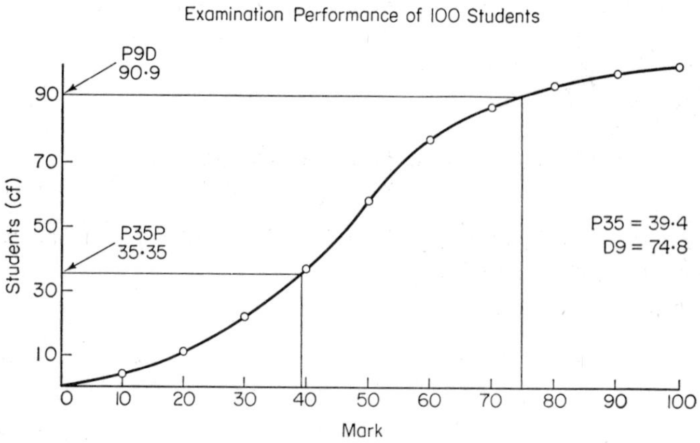

Fig. 5.—Ogive used to find a decile and percentile

The quartile deviation sometimes called the semi-interquartile range is found from the formula $\frac{Q3-Q1}{2}$. Thus if, by either the drawing method or the arithmetic method we found that the lower quartile was 50 and the upper quartile 160 then:

$$QD = \frac{Q3-Q1}{2} = \frac{160-50}{2} = 55$$

This measure tells us what the range of values in the Distribution would be if extreme cases at either end were ignored. It is particularly useful where a few particularly large or particularly small readings would otherwise give a distorted picture of the range.

6

Exercises (First Set)

The past examination questions in this chapter cover the work done in Chapters 1 to 5.

1A INCOME DISTRIBUTION BY AGE OF SAMPLE OF PROFESSIONAL ENGINEERS IN THE UNITED KINGDOM. 1962/3 (EXTRACT)

Income p.a.	*Aged 55 to 70*	*Total All Ages*
Under £800	5	9
£800-	18	86
£1200-	75	969
£1600-	96	1267
£2000-	115	758
£2400-	91	451
£2800-	72	245
£3200 and over	198	507

Draw histograms of the data in the two columns above and estimate the modes. Compare and contrast your results.

1B U.K. DRY CARGO MERCHANT VESSELS OF 500 GROSS TONS AND OVER 31.12.62

Age	*Number of Vessels*
Under 5 years	519
5 years and under 10 years	409
10 ,, 15 ,,	389
15 ,, 20 ,,	365
20 ,, 25 ,,	125
25 years and over	94
TOTAL	1901

From this table obtain by calculation the median age of vessels and the quartile deviation. What particular advantages do these measures have in connection with the distribution.

1C LENGTH OF LIFE IN HOURS OF A TYPE OF LIGHT BULB

592	739	558	686	554	747	803	629	725	627
644	644	534	742	629	552	693	469	625	586
731	558	622	594	528	697	373	691	702	581
648	622	824	673	566	566	684	564	663	657
573	483	666	613	614	614	876	683	577	729
634	633	569	722	640	640	670	609	503	607
575	768	519	643	684	658	563	605	561	624
548	655	656	603	603	777	441	651	641	541
653	622	713	637	647	611	667	594	611	646
658	709	639	661	661	794	762	675	666	659

Illustrate the above table on an ogive and hence find an estimate of the median. Check your answer by calculating the median by the arithmetic method (group by intervals of 50).

1D The following table shows the daily expenditure on the return journey to work of workers with a certain firm. Represent the distribution on an ogive and hence find the median expenditure and quartile deviation. If proposals to give some form of grant to those paying over 15/- per day were being considered by the firm, calculate the percentage of workers who would be affected and check the calculation by reference to your graph.

Expenditure	*Percentage of Workers*
Up to 6d.	8
6d. to 2/5d.	12
2/6d. to 4/5d.	15
4/6d. to 6/5d.	12
6/6d. to 8/5d.	10
8/6d. to 10/5d.	13
10/6d. to 12/5d.	10
12/6d. to 14/5d.	10
14/6d. to 16/5d.	5
16/6d. to 18/5d.	2
18/6d. to 20/5d.	1
20/6d. and over	2

1E DISTRIBUTION OF ESTATES ACCORDING TO VALUE

Value of estate in £000	*Number of estates*
0 but less than 2	1
2 ,, 4	5
4 ,, 6	16
6 ,, 8	25
8 ,, 10	12
10 ,, 12	9
12 ,, 14	17
14 ,, 16	35
16 ,, 18	35
18 ,, 20	18
20 ,, 22	30

From the above table draw a cumulative frequency curve and, from your curve obtain the answers to the following:

(*a*) What is the value of the Median?

(*b*) What is the quartile deviation?

(*c*) For the purposes of a certain statistical survey it is necessary to ascertain the range of values possessed by the first 40 per cent of the estates. What is the range? What name do we give to the measure which was used to discover this range?

7

Arithmetic Mean

The Arithmetic Mean is the third of the three common types of average used in Statistics. It is the normal type of average which we find by *adding all the numbers together and dividing by however many there are.* It is thus the best known average but it uses every value in the distribution and can therefore be distorted if there are extreme values at either end of the distribution.

It would clearly be inconvenient if we had to add a couple of hundred numbers in a distribution in order to find the Mean so we have a quicker method for finding the measure when we are given a frequency distribution.

(1) Set out the groups and frequencies without cumulation in a table and add a deviation column by putting a nought against the largest frequency; −1, −2, −3, etc. upwards from there to cover any groups above in the table; +1, +2, +3, etc. downwards from there to cover any groups below.

(2) Construct a frequency×deviation column (fd) by multiplying each frequency across by its deviation.

(3) Total the f and fd columns and thus find the arithmetic mean by the formula:

$$\text{CENTRE OF THE GROUP WITH ZERO DEVIATION} + \frac{\text{SUM OF THE FD COLUMN}}{\text{SUM OF THE F COLUMN}} \times \text{GROUP INTERVAL}$$

In the following example the ages of a hundred men are given in a frequency distribution and the arithmetic mean is calculated by the method outlined.

EXAMPLE

Age	*Number of Men* (*f*)	*Deviation* (*d*)	*Freq.*×*Dev.* (*fd*)
15-25	5	−2	−10
25-35	17	−1	−17
35-45	37	0	0
45-55	26	+1	26
55-65	15	+2	30
	100		56−27=29

The group of zero deviation ranges from 35 to 45 so the centre of it is 40. The sum of the f column is 100 and that of the fd column 29. A new group starts every ten years, so that is the Interval; thus:

$$A = 40 + \frac{29 \times 10}{100}$$

$$= 42{\cdot}9 \text{ years}$$

8

Standard Deviation

The Standard Deviation tells us *how far either side of the Arithmetic Mean most of the readings will lie.* Thus if the Mean mark in a Statistics test was 13 and the Standard Deviation was 4, we can assume that most of the class got between 9 and 17 marks—four either side of thirteen.

To find this measure we proceed as follows:

(1) Set out the group, frequency, deviation and frequency times deviation columns as we did when finding the Arithmetic Mean.

(2) Add a fifth column (frequency times deviation squared: fd^2) by multiplying the last two columns (d and fd) together on each horizontal line. Total this column.

(3) Calculate the Standard Deviation using the formula:

$$\text{GROUP INTERVAL}\sqrt{\frac{\text{SUM OF THE FD}^2\text{ COLUMN}}{\text{SUM OF THE F COLUMN}}-\left(\frac{\text{SUM OF THE FD COLUMN}}{\text{SUM OF THE F COLUMN}}\right)^2}$$

To show this further calculation clearly let us use the example from which we have already found the Arithmetic Mean:

Age	*f*	*d*	*fd*	*fd*²	
15-25	5	−2	−10	20	(−2×−10=+20 etc.)
25-35	17	−1	−17	17	
35-45	37	0	0	0	
45-55	26	1	26	26	
55-65	15	2	30	60	
	100		29	123	

$$\text{so SD} = 10\sqrt{\frac{123}{100}-\left(\frac{29}{100}\right)^2} = 10{\cdot}7$$

So if the Mean is 42·9 most readings will lie within 10·7 either side of that figure—i.e. between 32·2 and 53·6.

The Variance is a name given to the *square of the Standard Deviation.* It is useful because variances can be added and if two distributions had variances x and y, the variance of the two distributions combined would be x + y.

In our example the Standard Deviation came to 10·7 so the variance would be 114·5.

THE COEFFICIENT OF VARIATION

We are sometimes asked to compare the variability between two sets of figures. In such a case we use the coefficient of variation and the set with the higher coefficient will vary the most. The formula here is:

$$\frac{\text{STANDARD DEVIATION}}{\text{ARITHMETIC MEAN}} \times \frac{100}{1}$$

So using the Mean and Deviation already obtained in our example we get:

$$\frac{10{\cdot}7}{42{\cdot}9} \times \frac{100}{1} = 24{\cdot}9 = \text{COEFFICIENT OF VARIATION}$$

9

Normal Curve Areas

In the last chapter we said that the Standard Deviation indicated the limits either side of the Arithmetic Mean within which most of the readings will lie so that if the average height of a group of men is 5 ft. 10 in. and the Standard Deviation is an inch, then most men will lie between 5 ft. 9 in. and 5 ft. 11 in. Now it can be proved that the normal percentage of men who will lie within one Standard Deviation either side of the Arithmetic Mean is $66\frac{2}{3}$.

For greater accuracy it can similarly be proved that the normal percentage of any population who will lie within two Standard Deviations either side of the Arithmetic Mean is 95 and the percentage who will lie within three Standard Deviations either side of the Arithmetic Mean is $99\frac{3}{4}$. These are known as the normal curve areas.

Thus further observations on the heights of the men in our examples would show that 95 per cent lie within 2SD of the AM or within 2 × One Inch of 5 ft. 10 in. (5 ft. 8 in. to 6 ft.) and $99\frac{3}{4}$ per cent—which virtually means all of them can be taken as such—lie within 3SD of the AM or within 3 × One Inch of 5 ft. 10 in. (5 ft. 7 in. to 6ft. 1 in.).

EXAMPLE

The Mean height of a normal group of 1,000 plants is 20 in. and the Standard Deviation of the group is 2 in., state between what limits the heights of the plants probably lie. State probable number of plants whose heights will lie between 18 in. and 22 in.

All the plants (we ignore the odd $\frac{1}{4}$ per cent) will lie within three Standard Deviations of the Mean. In this case that means six inches either side of twenty inches or between 14 in. and 26 in. The number of plants which will lie between 18 in. and 22 in. is the number which will lie within one SD either side of the Mean which is $66\frac{2}{3}$ per cent or 667 plants to the nearest.

Other possibilities

If the question had asked for the percentage of plants between 16 in. and 24 in. this would mean 2SD either side of the Mean or 95 per cent.

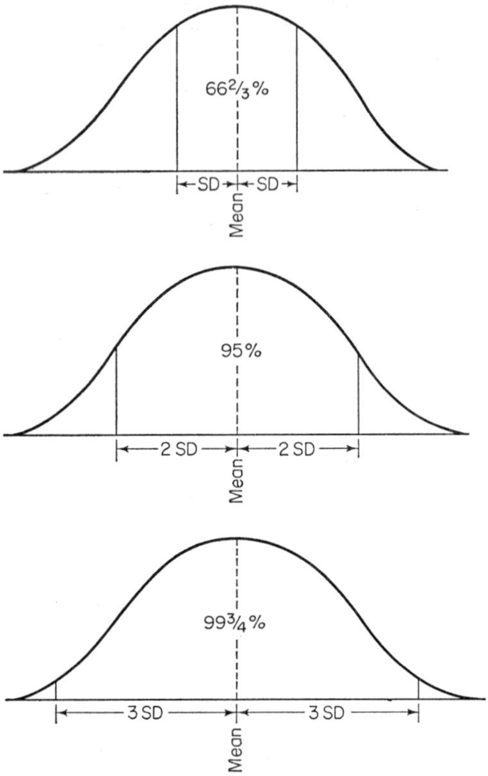

Fig. 6.—Normal curve areas

It is even possible to request something like 'the number of plants whose probable heights would be between 16 in. and 22 in.' In such a case we clearly need the number of plants within two Standard Deviations one side of the Mean and one Standard Deviation the other side of the Mean. Now if the number within 2SD either side of the Mean is 95 per cent., in a normal distribution this would mean that $47\frac{1}{2}$ per cent lay on each side of the Mean figure. Likewise if the number within 1SD either side of the Mean is $66\frac{2}{3}$ per cent, it follows that $33\frac{1}{3}$ per cent probably lay on each side of the Mean figure. So in a case like this we would combine the two halves and say: '16 in. is two Standard Deviations below the Arithmetic Mean and the number of plants within that normal area is $47\frac{1}{2}$ per cent. 22 in. is one Standard Deviation above the Arithmetic Mean and the number of plants within that normal area is $33\frac{1}{3}$ per cent.

Adding we find that the percentage which will probably lie between 16 in. and 22 in. is $47\frac{1}{2} + 33\frac{1}{3}$ per cent $= 80\frac{5}{6}$ per cent which is 808 plants to the nearest.'

10

Ratio Scale Graphs

When we are more concerned with the rate of growth of a set of figures than their progression, we illustrate them by means of a ratio scale graph or semi-logarithmic chart. The second name is derived from the paper which is normally used for such diagrams. Semi-logarithmic paper has normal horizontal scales but has logarithmic vertical scales. The effect of this type of ruling is that any number plotted on the vertical axis will in fact have its logarithm plotted instead. From this it follows that if the special paper is not available we can obtain the same effect by looking up the logs of the numbers which we wish to plot on the vertical axis and plotting the logs instead.

However we shall presume that the special paper is available. Horizontal readings are plotted in the normal way but to plot the vertical readings the following rules must be observed.

(1) In labelling the axis we do not start at 0 but at 1 or 100 or 1,000, etc.

(2) Care should be taken to label such figures against the darker printed horizontal lines as these are unevenly spaced along the axis. On some rulings of the paper feint guide numbers appear to assist in this.

(3) At the end of each cycle of ten vertical numbers, the group interval is multiplied by ten so that if we start at 1 we number to 10 and then go 20, 30, 40, etc. to 100 then 200, 300, 400, etc. to 1,000 and so on for as many cycles as there are on the paper. This gives us a high range of values which can be plotted on this paper and the fact must be kept in mind when choosing how many cycles to have on the paper. The semi-logarithmic paper is available with any number of vertical cycles up to 10 but one, two and three cycle are usually used and the two cycle is the commonest ruling.

EXAMPLE

The table shows the value of Exports in £000 of two firms each year for the last seven years. Draw a ratio scale graph to show the changes in the exports.

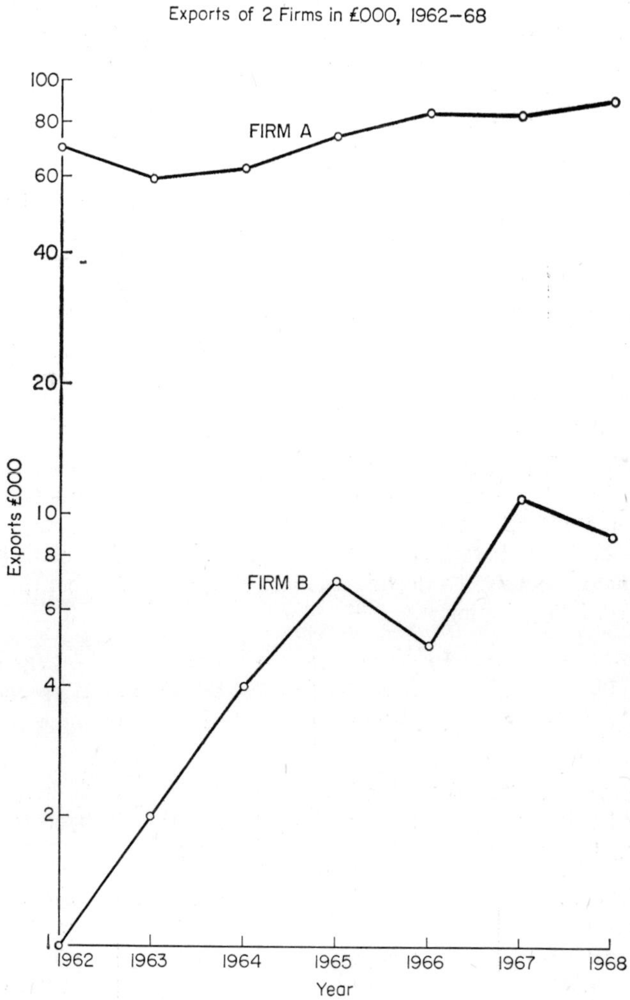

Fig. 7.—Ratio scale graph

Year	*Firm A*	*Firm B*
1962	70	1
1963	60	2
1964	63	4
1965	75	7
1966	85	5
1967	84	11
1968	91	9

In this example we have to cater for a vertical range from 1 to 91 which indicates the use of two cycle paper so that the first cycle can rise from 1 to 10 and the second from 10 to 100. The vertical axis is then labelled accordingly using the guide numbers to help. The horizontal axis is plotted normally and the two lines then filled in. The graph must then be fully labelled and the actual lines differentiated in some way—either by labelling them or by completing them in different colours. *See* Figure 7.

DIFFERING UNITS ON VERTICAL AXIS

It is in order, when drawing ratio scale graphs, to show two sets of figures with different units and different ranges of figures. We could show numbers of houses in hundreds and value of them in millions of pounds. To do this one can either double-label the left-hand vertical axis thus:

Houses	*Value*
00	£M
	10
	9 etc.

Alternatively the right-hand vertical margin can be used and labelled to show one of the sets of figures and the left-hand margin to show the other thus:

Houses		*Value*
00		£M
10		100
9, etc.		90, etc.

The latter method is preferable when the two sets of numbers are different also—as here when one rises to 10 and the other to 100.

11

Other Pictorial Methods

In earlier chapters we have dealt with the block diagram or histogram, the ogive which is a normal type of graph and the ratio scale graph or semi-logarithmic chart which is a special type of graph. Of the various other pictorial methods available to us, the following are the most important and we should know how and when to use them.

PIE DIAGRAM

This consists of a circle divided into segments like slices of a pie in accordance with the size of the figure involved. It is thus useful when we wish to show the relationship of a part to a whole and we might use it, for example, to show how much of the government's taxation income comes from different sources. It would not, on the other hand, be appropriate to compare income tax revenue in different countries for little would be gained by totalling such

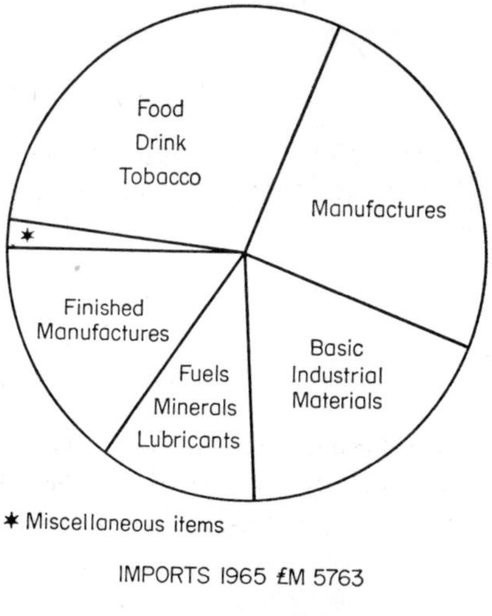

Fig. 8.—Pie diagram

revenues. In examination questions we are required to draw these diagrams with a great degree of accuracy by one of two methods of which the latter is the most common.

(*a*) Special paper can be obtained upon which a printed circle is divided by feint rules into one hundred segments. Taking each figure in our table we calculate how many segments of the hundred are appropriate to that figure using the formula

$$\frac{\text{INDIVIDUAL FIGURE FROM THE TABLE}}{\text{TOTAL OF THE FIGURES IN THE TABLE}} \times \frac{100}{1}$$

It is as well to work out how many segments are taken up by all the figures in the table before proceeding to mark off the sections and shade or colour them. In this way one can check that all the figures take up one hundred segments—no more and no less—and thus avoid errors.

(*b*) Where no special paper is provided a suitable circle must be drawn with compasses and divided into sections by means of a protractor. To do this the figures need to be transformed into three hundred and sixtieths of the total so that the correct number of degrees in the circle are allocated to each. Here the formula will be:

$$\frac{\text{INDIVIDUAL FIGURE FROM THE TABLE}}{\text{TOTAL OF THE FIGURES IN THE TABLE}} \times \frac{360}{1}$$

Once again it is as well to check that the total of degrees from all the figures is 360 before proceeding to split up the circle.

PICTOGRAM

A small recurring picture is used to produce a visual image of the figures. If, for example, a town had 12,000 television licences in

Fig. 10.—Pictogram

1958 and 16,000 in 1968 we might show twelve little T.V. aerials against the first year and sixteen against the second year under a suitable title. The important points to remember here are that there must be a key to show what each symbol means—in this case one aerial = a thousand licences—and the symbol must be simple to draw or too much valuable examination time will be spend in reproducing it. It should nevertheless be appropriate so an aerial seems the simplest item here—easier than a small T.V. set.

BAR CHART

This type of diagram can be used to illustrate many types of information but is particularly useful when illustrating how a certain whole figure varies in different countries of different situations. In such a case it is simpler than drawing a number of pie diagrams and is just as effective. In the diagram this method is used to show Taxation as a percentage of National Output and how the percentage varies in six different countries.

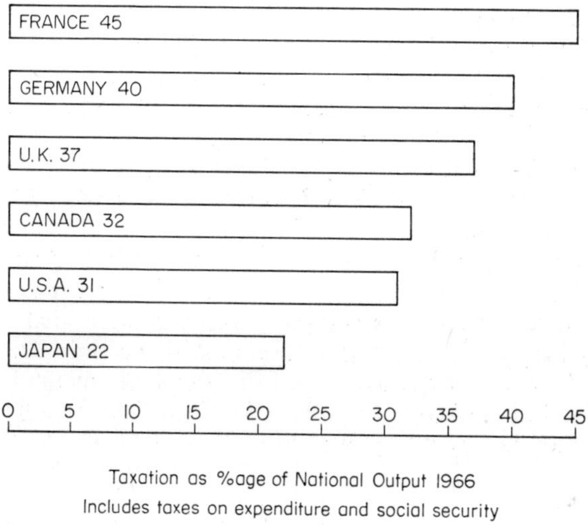

Fig. 9.—Bar chart

COIN DIAGRAMS

When financial figures are involved a drawing of piles of coins of appropriately differing heights is popular. At the top of each pile a coin can be fragmented to show fractional readings. Thus if a charity wishes to illustrate how it proposes to use an additional income of 3/2d. in the pound, a pile of eighteen coins making up the 1/6d. to be spent on education, a pile of twelve coins making up the 1/- to be spent on old people and a pile of eight coins making up the 'other works' allocation would illustrate the division very well and this is shown in the diagram. A suitable key and clear labelling must, of course, be added.

A Charity found its income had increased by 3s 2d in the £
The diagram shows how it proposed to spend the money

Fig. 11.—Coin diagram

MISCELLANEA

Apart from these different methods of illustration, we can show more than one set of figures on a graph by different lines—possibly in different colours—and we can even show that the original line was from figures which were a total of two or more earlier figures by shading in areas between the different lines as shown in the diagram by splitting each block into component parts each suitably shaded or coloured.

One often sees in books apt diagrams which follow none of the above rules but are a compound of several rules or even a separate entity in themselves. Thus diagram questions often give scope for ingenuity because very often a section requiring certain figures to be illustrated by one of the prescribed diagrammatic methods already outlined is followed by a second section in which some figures can be illustrated in any way you please.

12

Exercises (Second Set)

The past examination questions in this chapter cover the work done in Chapters 7 to 11.

2A A load of timber consists of planks of the following lengths

Length (ft.)	8-	9-	10-	11-	12-	13-	14-	15-	16-
No. of planks	1	3	6	11	15	12	9	6	3

NOTE: 8- means 8 ft. and over but less than 9 ft.
9- means 9 ft. and over but less than 10 ft., etc.
16- means 16 ft. and over.

Find the Mean length of the planks and the Standard Deviation.

If this load of timber can be considered as a random sample from a large cargo which has the normal distribution, make an estimate of the length of the longest plank you would expect to encounter.

2B U.K. MERCHANT VESSELS OF 500 GROSS TONS AND OVER ON 31.12.61

Age	*Number (Passenger)*	*Number (Cargo)*
Under 5 years	29	519
5 years and under 10 years	32	409
10 ,, 15 ,,	85	389
15 ,, 20 ,,	19	365
20 ,, 25 ,,	37	125
25 years and over	62	94
TOTAL	264	1901

Find the average age and the Standard Deviation of passenger vessels from the above table giving your answers correct to two places of decimals. Briefly state the value of such measurements in statistical analysis.

2C NUMBER OF GOODS VEHICLES IN GREAT BRITAIN IN 1961

Size (Unladen Weight)	*Number (Thousands)*
Under 12 cwt.	62
12 cwt. and under 1 ton	514
1 ton and under 2 tons	321
2 ,, 3 ,,	293
3 ,, 4 ,,	145
4 ,, 5 ,,	52
5 ,, 6 ,,	20
6 ,, 7 ,,	10
7 ,, 8 ,,	11
Over 8 tons	12
TOTAL	1440

From the above table calculate the mean size of vehicle and the Standard Deviation.

2D In a town of two wards the family size is shown in the table below.

(i) Calculate the Arithmetic Mean for Ward B and then both the Arithmetic Mean and Standard Deviation for Ward A.

(ii) Using the results already obtained, calculate an Arithmetic Mean for the town as a whole.

	People in Family						
Ward	1	2	3	4	5	6	7 & over
A	35	77	56	40	21	14	10
B	30	70	42	26	13	8	2

2E The following table shows the number of 'O' level subjects undertaken by 16-year-old Secondary School Pupils sitting the papers of a certain G.C.E. Board in 1968.

Number of Subjects	*Candidates in Hundreds*
1	45
2	32
3	36
4	47
5	73
6	103
7	133
8	132
9	51
10	8
11	1
	661

Calculate from the figures the Arithmetic Mean and Standard Deviation. Do you consider it would be appropriate to calculate the Coefficient of Variation here? Give your reasons.

2F The mean height of a normal group of 1,000 men was 5ft. 9 in. and the Standard Deviation was 2 in. State between what limits the probable heights of the men lie. How many men will be between 5 ft. 5 in. and 5 ft. 11 in. in height?

2G Total production of motor spirit in a certain country is given for various years in the table.

Year	1952	1954	1956	1957	1958	1959	1960
Production	267	334	635	700	919	1025	1143

m. Metric tonnes

On normal graph paper construct a ratio graph with semi-logarithmic scales to illustrate these figures.

Assuming that the trend of 1958, 1959 and 1960 continues into 1961, use your graph to estimate the total production of motor spirit in the country in 1961.

2H Using two cycle semi-logarithmic graph paper, construct a ratio graph from the following figures. What are the advantages of such graphs over those on arithmetic scales? How can one construct such a graph if the special paper is not available?

NEW FACTORY BUILDINGS IN RURITANIA

Year and Quarter		*Total*	*Number in Development Areas*
1964	i	264	92
	ii	267	122
	iii	215	48
	iv	276	90
1965	i	256	58
	ii	284	65
	iii	218	48
	iv	264	48
1966	i	241	59
	ii	248	65
	iii	271	46
	iv	179	51

2J State the main purposes of representing numerical data in pictorial or diagrammatic form.

Three types of diagram often used are (*a*) the Bar Charts, (*b*) the Pie Chart, and (*c*) the Pictogram. Illustrate these diagrams by suitable sketches. Discuss the main advantages and disadvantages of these charts, and give an example of the type of data which could be satisfactorily represented by each.

2K The following table shows the amount of personal income and expenditure for the U.K. in 1961 at current prices:

Income	*£m*	*Expenditure*	*£m*
Insurance benefits etc.	1796	National Insurance	1071
Rent, Dividends, Interest	2561	Savings	2087
From self employment	2075	Consumer spending	17302
Employers' contributions to Insurance & Pensions	1131	Taxes on Income	2230
Armed Forces Pay	387		
Wages and Salaries	14740		
	22690		22690

Present this information diagrammatically using:

(*a*) pie diagrams of two inches radius and

(*b*) any other form of diagram.

Give reasons for the type of diagram chosen in (*b*)

13

Approximation and Error

Statistics is not an exact science but a science of approximation. The basic idea in many cases is to find the best possible guess. Consequently there will often be a difference between the true figure and our estimate of it. The difference between the true figure and our approximation is the absolute error. If we express this as a percentage using the formula:

$$\frac{\text{ABSOLUTE ERROR}}{\text{TRUE FIGURE}} \times \frac{100}{1}$$

we have calculated the relative error. Let us suppose that the true figure in a certain calculation was 200 and our estimate was 240. Then the difference is 40 and that is the absolute error. The relative error from the above formula is $\frac{40}{200} \times \frac{100}{1} = 20$

A third type of error is the biased error where *every* estimate is above the true figure or *every* estimate is below the true figure instead of some estimates being above it and others below. Thus if every child in a class was given an age of the number of years old last birthday, then every estimate would be below the true age and we would have a number of biased errors ranging in size from a few minutes to just short of one year.

Now let us deal with the various ways in which figures can be approximated.

(1) TO THE NEAREST ? If 32, 47 and 566 were all taken to the nearest ten the estimates would be 30, 50 and 570.

(2) TO THE NEAREST ? BELOW If the same figures were taken to the nearest ten below, the estimates would be 30, 40 and 560.

(3) TO THE NEAREST ? ABOVE The same figures to the nearest ten above would yield estimates of 40, 50 and 570.

(4) TO ? SIGNIFICANT FIGURES 367, 5890, 34441 taken to two significant figures would produce estimates of 370, 5900 and 34000.

(5) TO ? PLACES OF DECIMALS 3·57, 56·344 and 7656·67 each taken to one place of decimals would produce estimates of 3·6, 56·3 and 7656·7.

(6) TO WITHIN ? EITHER WAY The numbers 35 and 47 to within 2 either way would yield estimates of 33 to 37 and 45 to 49 respectively.

EXAMPLES

It is very difficult to give hard and fast rules for dealing with problems on error. In other chapters rules can be given for the problems involved are somewhat stereotyped. Here however they are like riders in Geometry—every one is different. A basic rule can be given however. *Work out the various limits possible within the figures given and combine them suitably to produce the answers required.*

TYPE 1

If the male population of four towns is given as 20,000+46,000+12,000+29,000 and the unemployed to the nearest ten in each case were 480, 920, 450 and 630 we could be asked to find the upper and lower limits for the total unemployed and the relative error expressed as a percentage. To do this we add the lower possibilities—475, 915, 445 and 625 thus getting the lowest possible total and then add the higher possibilities 485, 925, 455, 635 thus getting the highest possible total. Then using the formula stated:

$$\text{LOWEST RELATIVE ERROR} = \frac{\text{LOWEST TOTAL UNEMPLOYED}}{\text{TOTAL POP. OF 4 TOWNS}} \times \frac{100}{1}$$

$$\text{HIGHEST RELATIVE ERROR} = \frac{\text{HIGHEST TOTAL UNEMPLOYED}}{\text{TOTAL POP. OF 4 TOWNS}} \times \frac{100}{1}$$

Care must be taken in combining the various limits together. If we had highest and lowest number of gallons of petrol used by a bus company in a given week and the highest and lowest number of miles run in the same corresponding weeks, the greatest miles per gallon possible would be highest number of miles travelled divided by lowest possible fuel used and the lowest miles per gallon would be least number of miles travelled divided by the most possible fuel used.

So we must not always think of combining highest with highest or lowest with lowest. The method will vary with the sense of the question.

TYPE 2

If a room measured 12 ft. by 10 ft. to within two feet either way and the walls were 10 ft. high to the nearest foot we might be asked the surface area of the walls for decorating purposes after subtracting a 6 ft. by 3 ft. door and a 4 ft. square window. From this we find the greatest possible sizes of the walls are two at 14 ft. by 10 ft. 6 in. and two at 12 ft. by 10 ft. 6 in. less 18 square feet for the door and 16 for the window.

The least possibilities are two walls at 10 ft. by 9 ft. 6 in. and two at 8 ft. by 9 ft. 6 in. less the 18 and 16 square feet from the totals in square feet.

Most of the problems fall roughly into one of these two main types but the student is urged to obtain as much practice from past papers as possible here for thought is always more important than memory in solving error problems.

14

Survey Work and Sampling

Most of the figures we use in our calculations come, either directly or indirectly, from some kind of statistical survey taken previously.

SURVEY WORK

The four main ways of carrying out such a survey are as follows:

(1) DIRECT OBSERVATION

In some cases actual observation of all cases is possible. One could for instance easily count directly all the cars passing a certain set of traffic lights in one direction.

(2) INTERVIEW

In this case teams of interviewers are sent out with instructions on who to question and forms telling them what to ask. This method is often used to find out how many people can tell one brand of a product from another or how many people actually use 'X' soap powder.

(3) QUESTIONNAIRE BY POST

Where the people to be surveyed are very scattered one can send them a form to fill in by post. A society with two thousand members all over the country might use this method to build up some statistical information about the make-up of its membership.

All these methods have disadvantages which introduce error into the results. Direct observation can be unduly expensive. Interviews can yield false information through misunderstanding on either part or even a definite wish to mislead on the part of the interviewee. Questionnaires by post are the least satisfactory of all because so many people will fail to return the forms and those who do may not provide a representative cover of the areas to be researched.

(4) ABSTRACT FROM PUBLISHED STATISTICS

This means taking the figures from published surveys by the

Government and other bodies. This method can have any of the above disadvantages hidden below the surface.

It is therefore necessary to weigh up all the pros and cons of each method in the given situation and choose the method which seems best in the particular case.

PILOT SURVEY

Before leaving the subject of survey work it must be mentioned that sometimes a small survey is made prior to the major work in order to test the forms or to give the interviewers experience after which a meeting can be held to iron out any difficulties encountered. Such a trial run is known as a pilot survey.

SAMPLING

Having chosen the method to be used we must now decide what people we are going to ask out of all the people available to ask. If we wanted to find out what kind of razor men in a certain town prefer we could hardly ask every man in the town. We must think of some way of taking representative opinions. This is known as taking a sample of the population. There are three main types of samples:

(1) THE RANDOM SAMPLE

Here we decide on a number of people and choose them at random. We might decide to interview a hundred men and go out and talk to the first hundred men we see. The snag here is that we will probably not get a representative sample of all ages and income groups by taking the first hundred men we see.

(2) THE STRATIFIED SAMPLE

In this case we decide on the number to be interviewed and divide that number into strata. Thus we might divide the hundred men into ten sub-divisions by age or by income or by the area in which they live. This method is better but more expensive.

(3) THE MULTI-STAGE SAMPLE

Here we break down the total even further by bringing two or more variables into the matter. One hundred men might then be divided into five age groups and each of the five age groups into five income categories thus getting twenty-five sections overall. This method is the best but is also the most expensive and time-consuming.

Two less-common types of sample are the *quota sample* where each interviewer is sent to an area and asked to interview everyone until he reaches a certain figure and then to stop as that figure is his 'quota'. Thus we might send five men to different parts of the town and tell each to interview the first twenty men and then stop. The final method is *cluster sampling* where we take one type of person as representing the whole. Thus we might take a tall dark-haired man as representative and tell our staff to interview one hundred tall dark-haired men.

EXAMPLES Examination questions on survey work will usually either pose some questions on the elementary theory of the work as already outlined or will give the gist of some survey work which you are to imagine you are going to carry out and ask you to write notes on how you would do it. Like error questions these are based on the steps indicated beow.

(1) Decide on the method to be used, the sample to be taken and the time at which it will be appropriate to work.

(2) Draw up any forms and instructions necessary for the operation. Decide on the number and type of interviewers required and brief them.

(3) Try to find out what possible sources or error or bias exist in the work and endeavour to minimize these.

(4) Carry out the work.

(5) Collect and sort the data. Draw up conclusions and recommendations and pass these on to the appropriate body.

Examination questions are also set in which a rather more complex type of form has to be drawn up and that is all. Provision has to be made for all possible sub-totals and the student is strongly advised to work these out fully in rough before starting to rule the final form.

15

Sources of Statistics

There are a great many sources of Statistics both government-sponsored and privately sponsored. Questions on sources are most unpopular with students at present because the work involved in adequately covering this topic for one possible question is out of all relation to that involved in preparing for any other question the paper is likely to pose. As an absolute minimum the student should work through the following publications and learn the notes which follow together with any other details he can.

(1) ANNUAL ABSTRACT (C.S.O. Annually)

(*a*) Area and Climate.

(*b*) Population and vital statistics (births, deaths, etc.). Geographical Distribution, Marital Condition, etc.

(*c*) Social Conditions: Expenditure on National Insurance, Allowances, Health Service, Pensions. Houses occupied and unoccupied. Licensed premises, clubs and cinemas. Libraries. Number of electors. Police force statistics.

(*d*) Education: Schools, Colleges, Universities, Meals, Milk, Grants, Staff. Income and Expenditure.

(*e*) Labour: Distribution of manpower, Earnings, Union and Strike statistics.

(*f*) Production: Fuel and Power, Iron and Steel, Building and Construction, Manufactured Goods, Agriculture, Fisheries and food.

(*g*) Retail Distribution: No. of establishments, sales, turnover, etc.

(*h*) Transport and Communications: Road Vehicles, Railway stock, Aircraft. Mileage. Accidents to same. Ships and shipping movements. Postal, Telegraph and Telephone services.

(*i*) External Trade: Imports and Exports by class and quantity with index numbers.

(*j*) Overseas Finance: Balance of payments, overseas holdings, economic aid and foreign exchange rates.

(*k*) National Income and Expenditure: Personal, Government and Company.

(*l*) Home Finance: Income and Expenditure of Government. National Savings. Rateable values. Local Government Income and Expenditure. Debts.

(*m*) Banking and Insurance: Including Assurance, H.P., Number of Companies and insolvencies.

(*n*) Prices: Retail and Wholesale. Shipping Rates. Agriculture and Livestock. In most cases figures are given for the last ten years to enable trends and comparisons to be studied.

(2) MONTHLY DIGEST (C.S.O.)

A monthly version of the Annual Abstract with rather more detail and more recent figures. Similar tables to the above cover the five major topics of Population, Production, Trade, Labour and Prices.

(3) ECONOMIC TRENDS (C.S.O.) Monthly

An abstract of statistics on the five major topics are illustrated by graphs and diagrams. Each issue has special articles on topics of current economic interest and the March issue each year has, as a supplement, an essay on economic events of the past year which is most valuable.

(4) MINISTRY OF LABOUR GAZETTE (Ministry of Labour) Monthly

Statistics relating to employment, unemployment, unfilled vacancies, overtime and short-time working, earnings, hours of work, prices and industrial disputes.

In addition each edition has special articles relating to labour and details of any recent arbitration awards and statutory instruments.

(5) BOARD OF TRADE JOURNAL (Board of Trade) Weekly

Very detailed and up-to-date statistics on production, prices and trade under the following topics:

(*a*) Sales, work done, work in progress and stocks. Index of Industrial Production.

(*b*) Statistics on all sorts of industrial materials and manufactured goods.

(*c*) Distribution.

(*d*) Imports and Exports. Tariffs, Customs and regulations in overseas countries.

(*e*) Prices and Index Numbers of Prices.

(*f*) Passenger movement in and out of U.K.

(*g*) H.P. Outstanding Debt etc.
(*h*) Cinemas and films.

(6) NATIONAL INCOME BLUE BOOK (C.S.O.) Annually

General survey of national product and income and expenditure from census of production figures, central government accounts and balance of payments white papers. These varied sources make combination of figures difficult and the results can be said to be synthetic but the book is nevertheless the main source on G.N.P. and Income and Expenditure.

(7) CENSUS OF POPULATION

(Leading to various publications of parts and sections over a range of time following each Census. First Census was in 1801 then usually every ten years. Power to increase frequency to five years is held and a ten per cent sample survey was taken in 1966.)

In a full Census such as that of 1961 everyone answers questions on Name, Sex, Age, Marital Condition, Address, Place of Birth, Details of any dual nationality (plus a question on Welsh Language in Wales and Gaelic in Scotland).

Then a 10 per cent sample answer further questions on Occupation, Employer, Place of work, Hours worked, Previous Occupations, Education, Qualifications, Recent changes of address and the number of people in the house on the census night.

The amount of information subsequently published is both interesting and surprising. The findings are used by government, local authorities, industry and trade unions, universities, medical and social workers etc. to see present situations and plan ahead.

The Census of Production is obtained by similar means for similar purposes from industrial undertakings of all kinds.

(8) FINANCIAL STATISTICS (C.S.O./Bank of England) Monthly

A recent addition to the source list, this important publication covers:

(*a*) Central Government Finance.
(*b*) Local Authority borrowings.
(*c*) Public Corporation accounts.
(*d*) Banking Statistics and figures on Discount Houses, Savings Banks, H.P. Unit Trusts, Insurance and Building Societies.
(*e*) Income and Profit of Companies. Capital Issues.
(*f*) Interest Rates, Mortgage Rates, Security Prices and Yields.
(*g*) Gold and Foreign Currency Reserves, Balance of Payments, Exchange Rates.

16

Index Numbers

Index numbers are a convenient way of showing at a glance the *overall meaning of a number of changes in prices or other figures over a period of time.* The year at which our observations start is known as the base and each figure for that point of time is given an index of 100. The index for each commodity at the other point of time is calculated accordingly. This gives us Individual Price Indexes for each item. These can then be totalled and averaged to give an Average Price Index for all items.

On the opposite page three simple price changes are illustrated. You will see that the prices begin in 1946 which is thus our base. These prices are given an Index of 100. We always take 100 as the base (rather than say 0 or 1) so that we can cater for possible reductions too. The later prices for 1968 are then related to this figure and the fourth column contains Individual Prices Indexes for 1968 based on 1946. The said indexes are then totalled and averaged to give an Average Price Index for 1948 based on 1946.

Item	*1946 Price*	*Base Index*	*1968 Price*	*Individual Index* (to nearest whole number)
Bun	3d.	100	5d.	167
Ice Cream	3d.	100	6d.	200
Mineral Water	5d.	100	7d.	140
				507

Dividing by 3 we get AVERAGE PRICE INDEX=169

Sometimes we may wish to take a later point of time as the base and work backwards to an earlier point of time. Such a case illustrates the necessity of taking 100 as the base, for the earlier figures will probably be below 100. Here is our earlier example with the 1968 prices taken as 100 and those for 1946 worked out accordingly to the nearest whole number. An average index for 1946 based on 1968 has been added.

Item	*1946 Price*	*Individual Index*	*1968 Price*	*Base Index*
Bun	3d.	60	5d.	100
Ice Cream	3d.	50	6d.	100
Mineral Water	5d.	71	7d.	100
		181		

Dividing by 3

AVERAGE INDEX=60

In practice it is more usual to go to two places of decimals unless told to the contrary.

It is not usually accurate however to give equal emphasis to each price change. A rise in the price of rice in China would be very important to the people there whereas it would matter little here. A rise in the price of diamond rings would matter little to the average man whereas a similar percentage rise in the price of beer would matter a great deal. It is therefore usual to introduce weights which are a ratio of the importance of the various items compared to the others and are usually compiled from the last available figures of total production or consumption etc.

The individual index numbers are then multiplied by the individual weights of produce weighted index numbers and these are then totalled and divided by the sum of the weights to produce a weighted average index number. Here is our original example once again. This time weights have been added to produce weighted index numbers and a weighted average index number for 1968 based on 1946.

Item	*1946 Price*	*Base Index*	*1968 Price*	*Index*	*Weight*	*Ind.×Wt.*
Bun	3d.	100	5d.	167	3	501
Ice	3d.	100	6d.	200	1	200
Mineral	5d.	100	7d.	140	2	280
						981

We divide by the sum of the Weights (6) to get WT. AV. INDEX NO.=$163\frac{1}{2}$.

Examination questions often require some gaps to be completed in a table like the one we have just considered.

Item	*Unweighted Index*	*Weight*	
Bun	167	3	WEIGHTED
Ice	X	1	AVERAGE INDEX
Mineral	140	2	IS $163\frac{1}{2}$

If we were asked to find X—the unweighted index for Ice Cream—from the above we know that the individual index numbers are multiplied by the individual weights and the results are totalled and divided by the sum of the weights to get the weighted average index number. In this case therefore

$$\frac{167\times3+1\times X+140\times2}{6}=163\tfrac{1}{2} \qquad \text{so } \frac{781+X}{6}=163\tfrac{1}{2}$$

$$\text{so } 781+X=163\tfrac{1}{2}\times6$$
$$\text{so } 781+X=981 \text{ and X is } 200.$$

This is a simple example of course but the questions are usually variations on that theme. From the rule

Total of the Individual Indexes multiplied by Individual Weights, divided by the sum of the Weights gives the Weighted Average Index Number

we work out a formula with one missing element and then find that element by deduction.

REBASING

When an index number becomes rather too high, it is usual to start again and rebase it at a hundred at the point of time to which the latest available figure relates. Subsequent changes are then measured from the new base year or time.

In such cases we are occasionally required to calculate, given the new Index at some point of time, what the old Index would have reached by then if it had remained in force. In such cases the formula we use for the calculation is:

LAST AVAILABLE FIGURE OF THE OLD INDEX + X% OF ITSELF

(where X is the percentage change recorded under the new Index)

EXAMPLE

The Index of a commodity in 1968 based on 1960 was 140. The previous Index based on 1950 had reached 130 when superseded. What would the old Index have reached in 1968 if it had remained in force?

The last available figure under the old index is 130 as stated and the percentage change under the new Index is 40. So the calculation becomes 130 plus 40 per cent of itself or:

$$130 + \left(\frac{40}{100} \times \frac{130}{1} \right) = 130 + 52 = 182 = \text{Answer}$$

17

Skewness

If a block diagram is drawn from a frequency distribution it will usually show low frequencies on the left, an increasing number towards the centre where there will be a peak and then a tailing off to the right. This is known as a normal distribution.

Various things can disturb this pattern. For example in Kirkby near Liverpool there is an exceptionally high number of young people in the population. A block diagram would therefore have its peak on the left rather than in the centre. A place like Bournemouth where there are many retired people would probably produce a histogram with a peak towards the right. Where the symmetry of the distribution is disturbed in this way, it is said to be skewed and we can measure the direction and degree of the skewness.

There are several ways of taking this measurement but two methods will be adequate for our use. If we are either given, or can more easily find the Mean and the Median we use what is known as the *Pearsonian Coefficient of Skewness* found from the formula:

$$\text{SK} = \frac{3\ (\text{MEAN MINUS MEDIAN})}{\text{STANDARD DEVIATION}}$$

An alternative formula can be used when we are given, or can more easily find the Mean and the Mode. It is:

$$\text{SK} = \frac{\text{MEAN MINUS MODE}}{\text{STANDARD DEVIATION}}$$

The case of the age groups in Kirkby where the peak is on the left is known as positive skewness and the case of Bournemouth where the peak may be on the right is known as negative skewness.

The range of possible values for the Coefficient of Skewness in theory is from -3 to $+3$ but in actual practice a high degree of skewness will come to plus or minus one or thereabouts. Thus the Kirkby case would have a positive value and then we pass through a coefficient of zero which would signify a normal distribution to negative values in cases like that postulated for Bournemouth.

EXAMPLE

Let us suppose that the Mean Age of the Kirkby population is 25, the Standard Deviation is 20, the Median is 24 and the Mode is 22. By the Pearsonian formula we get:

$$\text{SK} = \frac{3(25 - 24)}{20} = 0{\cdot}15$$

Using the other formula taking the Mode into account we have:

$$\text{SK} = \frac{25 - 22}{20} = 0{\cdot}15$$

In practice you will not usually compare both formulas but will just find the coefficient by the method most suitable in the case. If one did compare the use of both formulas the results would probably be close to each other but not identical.

18

Exercises (Third Set)

The past examination questions in this chapter cover the work done in Chapters 13 to 17.

3A Explain the term Statistical Error. Approximate the figures in the following table to:

(*a*) the nearest thousand,
(*b*) the nearest hundred,
(*c*) the next thousand below,
(*d*) two significant figures.

Calculate the absolute and relative errors of their totals.

DELIVERIES OF AGRICULTURAL MACHINERY 1962

Type of Machine	*Number*
Agricultural Tractors	201,952
Market Garden Tractors	36,609
Mowers	12,226
Balers	15,707
Milking Machines	7,194
Tractor Ploughs	17,013
Disc Harrowers	7,499

3B HOLDER HOLDINGS LIMITED

Name of Driver	*Miles Travelled 1963*	*Gallons of Petrol*
Smith	8,000	230
Jones	9,000	300
Robinson	12,000	480
Brown	6,000	400

The above table shows the miles travelled by the four drivers of a certain firm during 1963 and the amount of fuel they used.

(1) Assuming that the figures are exact give the average miles per gallon of each car to the nearest whole number.

(2) If the figures in the 'Miles Travelled' column are correct to the nearest thousand and the figures in the 'Gallons of Petrol' column are correct to the nearest ten, calculate the greatest and

least possible figures for the average fuel consumption in miles per gallon for the firm as a whole. These answers should also be calculated to the nearest whole number.

(3) Comment briefly on the figures obtained as answers to the first two parts of this question.

3C Your statistical unit is asked to conduct a traffic survey at a certain crossroads to see whether the phasing of the traffic lights is satisfactory. Describe how you would organize and carry out the survey and the criteria upon which you would base your report. What difficulties would you expect and how would you try to overcome them?

3D Define a random sample. A retailer of electrical goods, before opening a new branch, wished to know what goods were most popular in the district. He instructed three interviewers to request a 'random sample' of the population to answer a questionnaire. The three interviewers set about their jobs in different ways. The first visited every tenth person in the telephone directory. The second waited outside a large factory producing electrical goods and questioned people as they came out. The third visited every fifth house on a large housing estate. Discuss the extent to which each interviewer was likely to obtain the opinions of a 'random sample' of the population.

Suggest a method that might have been more successful than any of those employed, commenting on its particular advantages and disadvantages.

3E In tiling a floor, a builder estimated the length and breadth to be 30 ft. and 12 ft. respectively, each being subject to a possible error of one foot either way. He wanted to cover the floor with oblong tiles 12 ft. by 6 in. which would be laid with the longer side parallel to the length of the floor. The cost of these would be 4d. each with a possible error of 25 per cent. Find the greatest and least possible cost of the tiling and, in the case of the larger figure only, the relative error in the form of a percentage to the nearest whole number.

3F A man owns a piece of ground and he estimates that it measures 30 ft. by 36 ft. to within 3 ft. either way. He wishes to sow the ground with lawn seed but intends leaving in the middle a square with sides measuring exactly 6 ft. for a flower bed. The seed is thought to cost 2/1d. per square yard with a possible error of up to 4 per cent either way. When the seed is bought, any order to cover more than 100 square yards of ground is subject to a discount of one penny in the shilling with fractions of a shilling ignored. What was the man's estimate of the cost and what are the greatest and least possible costs of the project?

3G Write an account of the main statistical information contained in any ONE of the following official publications:

(i) The Ministry of Labour Gazette.
(ii) The National Income Blue Book.
(iii) The Board of Trade Journal.

What is the value of such information in the publication you have chosen?

3H Describe the nature of the statistical information concerning the U.K. as found in official publications on any *four* of the following topics:

(i) Index of Retail Prices.
(ii) Wholesale Prices.
(iii) Industrial Stoppages.
(iv) Production of Coal.
(v) The National Income.
(vi) Imports and Exports.

3J INDEX OF RETAIL PRICES—15 AUGUST 1961 (17 JANUARY 1956 = 100)

Group	*Index*	*Weight*
ALL ITEMS	115·7	1000
(1) Food	109·9	350
(2) Housing	138·4	87
(3) Fuel and Light	122·2	55
(4) Transport and Vehicles	123·4	68
(5) Miscellaneous Goods	125·8	59
(6) Services	126·6	58

The 'All Items' index is calculated for ten groups which include the six listed separately above. Calculate a single weighted average index number of retail prices for the four groups not mentioned above for 15 August 1961 based on 17 January 1956. If the Interim Index of Retail Prices for all items for 17 January 1956, (15 January 1952 = 100) was 115·8, estimate an index of retail prices for all items for 15 August 1961 based on 15 January 1952 and discuss any unsatisfactory features in the linking process used in your estimation.

3K Prices of five goods on 31.12.50 and in 1959 are given below. Using a weighted arithmetic mean of the relatives, calculate an index of prices for all these goods for 31.12.59 based on 31.12.50.

Class	*Weight*	*Price per lb. 1950*	*Price per lb. 1959*
1	10	1/3d.	2/1d.
2	12	3/4d.	5/0d.
3	20	2/6d.	2/1d.
4	6	4/2d.	3/10d.
5	12	2/0d.	2/4d.

3L What is the purpose of weights in index number construction? Calculate the two figures required to complete the spaces marked X and Y in the table below. The base year is 1950.

Commodity	*1955*	*1965*	*Weight*
A	130	173	7
B	125	Y	21
C	148	190	30
D	107	131	24
E	121	151	18
TOTAL	X	159·02	

If an earlier index based on 1930 had reached 170 when replaced in 1950, what would it have reached in 1965 if it had continued in force?

3M (NOTE: Pearson's Measure of Skewness will often follow a question involving calculation of the Arithmetic and Standard Deviation. The example below is thus concerned with work done in Chapters 7 and 8 as well as Chapter 17.)

The table below gives the number of earners found per family in a social survey which was carried out in a large town. Calculate the Mean and Standard Deviation. Given that the Mode is 1·7, find the Pearsonian Measure of Skewness and comment on the information provided by each of the three calculations.

Number of Earners in the Family	*Number of Families*
0	117
1	1086
2	616
3	287
4	126
5	47
6	14
7	4
8	1
	2298

19

Time Series

We have normally been concerned so far with statistics based on a number of readings on one theme gathered at one specific point of time. We must now discuss statistics where the same reading is given for a number of different points of time such as 'every year for the past 20 years'. Such a set of readings is called a time series and these studies enable us to see whether the figures are tending to increase or decrease. There is also a method of giving an estimate of future figures.

If the figures of a time series are plotted on a graph the result will usually be an oscillating curve. The length of time from peak to peak will be reasonably constant though the height of each peak may vary. This length of time is known as the cycle and before we deal with a time series we must determine the cycle.

Sometimes the cycle is clearly given by such a phrase as, 'Using a five year cycle calculate . . .' or 'using a six month moving average determine . . .' (the moving average is the next step so this means the same thing). If there is no such phrase and the figures are in months or quarters, we are entitled to assume that the cycle is one year unless we are told to the contrary. It is only on rare occasions when the figures are in days, weeks or years and no other clue is given that we have to actually calculate the cycle. The method is as follows:

(1) Look for a high figure near the start of the table which could well be a peak figure.

(2) Look at every third figure thereafter and see if a three figure cycle will work throughout the table. The last figure of each cycle must be higher than any other figure in the cycle. As soon as this is not so, the pattern breaks down and the supposed cycle is wrong.

(3) Try next every fourth figure after the first peak . . . then every fifth and so on until some pattern works throughout the table. The first pattern to work throughout will be the cycle.

EXAMPLE

Year	*Figure*
1946	116
1947	131
1948	83
1949	100
1950	101
1951	92
1952	114
1953	115
1954	117
1955	123
1956	110
1957	94
1958	100
1959	103
1960	130
1961	114
1962	123
1963	142
1964	126
1965	101
1966	124
1967	107
1968	129

(1) The first obvious peak here is 131—the figure for 1947.

(2) If we take every third figure thereafter the system breaks down in 1956 for the figure for 1955 is higher than that of 1956 so the last figure of that cycle is not the highest in that three so three will not work.

(3) Four years breaks down at once for the 1951 figure is not the highest of the first four—1948, 1949, 1950, and 1951.

It is only when we get to eight years that we find a pattern which will work throughout. The cycle ends are marked after each eighth year and you will find no higher figure earlier in any given eight. Eight years is therefore the cycle.

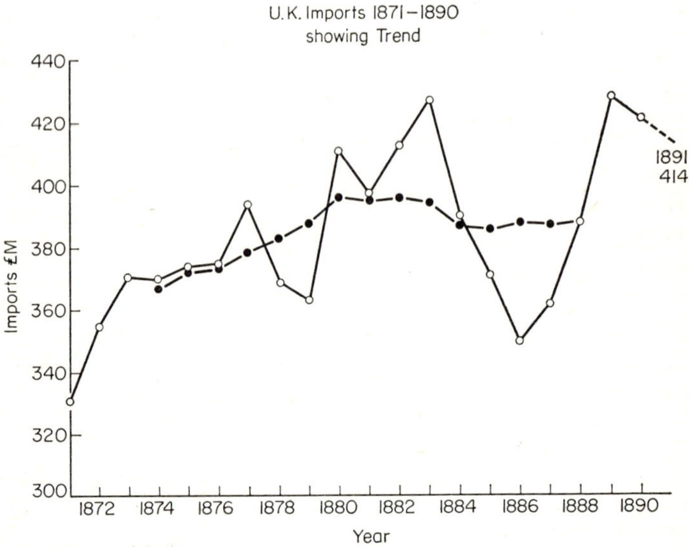

Fig. 12.—Time series showing originals and trend with former projected to produce estimate for next point of time

CALCULATIONS

Having determined the cycle one way or another we now work out the trend by the method of moving averages, The trend is, in fact, a set of readings which, if plotted on a graph, would produce a flattened-out curve showing whether the figures have a general tendency to increase or decrease.

Method 1. For use when the cycle is an odd number.

(*a*) Add together the first cycle of figures, average and enter the trend on the right.

(*b*) Move down one figure and repeat. Continue to foot of column.

EXAMPLE

Here is a set of figures where the cycle is given as 7 years so we add the first seven figures, then average them by dividing the total by seven. Then we move down one figure and continue until we reach the foot of the column:

Year	*Imports £M*		*Moving Average*			*Trend*
1871	331					——
1872	355					——
1873	371	2570				——
1874	370	7	2608			367·14
1875	374		7			372·57
1876	375			etc.		etc.
1877	394				etc.	etc.
1878	369					
etc.	etc.					

NOTE that we put each trend figure opposite the middle figure of the cycle and this means that we shall have no trend figures for the early years of the time series or for the last years either.

Method 2. For use where the cycle is an even number.

(*a*) Work out a total for the first cycle of figures and put it down without averaging. Move down one figure and get the next total. Continue to the foot of the column.

(*b*) Work out a grand total for the first two totals, then for the next two totals and so on to the foot of the column.

(*c*) Average each grand total by dividing by twice the cycle figure. These are the trend figures which should be entered in the trend column opposite the centre of the two totals involved in each.

EXAMPLE

Here is a set of figures in quarters so we can assume the cycle is a year which means four figures. So we total the first four figures, then the next four and so on. We add each pair of totals and then find the trend figures as averages of grand totals by dividing by eight.

Year	*Quarter*	*Figure*	*Fours*			*Eights*	*Trend*
1959	1	122					
	2	117	471				
	3	109		471		942	117·75
	4	123				945	118·25
1960	1	122			474 etc.	952	119
	2	120					etc.
	etc.						

ESTIMATING

When asked to estimate a figure for a period later than those given in the table:

(*a*) Look at the last three or four trend figures and make an intelligent estimate from them of what the next figure might have been.

(*b*) Multiply backwards to find what the relevant total or grand total must have been.

(*c*) From that deduce what the next original figure might have been. Continue until the figure wanted is found.

EXAMPLE

From the following table calculate a figure for the last quarter of 1961. (Only the bottom end of the table is reproduced.)

Year	*Quarter*		*4s.*	*8s.*	*Trend*
1960	4	129			120·25
1961	1	124			121·25
	2	121	485		121·62
	3	115	489	976	122·00

Here we might guess the next Trend could be 122·25. Multiplying by eight we would get the next eight as 978. That must have been made up of the last four (489) plus an unknown total which must also then have been 489. That four must have been made up of the last three original figures plus the unknown figure for the last quarter of 1961 which is the one we want. We thus find:

1961 4=124+121+115 all taken from 489

So the figure required must be 129

20

Seasonal Fluctuations

A man who sells rainwear or umbrellas can expect to do more trade in the Winter when the weather is bad than he will do in the summer. Likewise a purveyor of Mineral Waters can expect to do more trade in the Summer when the weather is hot than he will do in the cold conditions of winter. These are two rather obvious examples of seasonal variations and all trades experience them to a greater or lesser extent.

No time series study is really complete without a consideration of these factors and consequently many examination questions on time series require the calculation of the seasonal variations which means that we find the amount over or under a normal figure which the figure for a given season is expected to be. Alternatively you may be asked for seasonally adjusted figures which means that, after having found the seasonal variations, we have to take the appropriate amount away from, or add the appropriate amount to certain of the original figures which were given in the table.

Having found our trend by the method outlined in the previous chapter, the first thing to calculate is an initial variation column by subtracting each trend from the original figure it is placed opposite. Thus taking the figures in the second example of the last chapter we would get:

Year	*Quarter*	*Figure*	*Trend*	*Initial Variation* (Original-Trend)
1959	1	122	—	—
	2	117	—	—
	3	109	117·5	−8·5
	4	123	118·25	+4·75
1960	1	122	119	+3
	2	120	etc.	etc.
	etc.			

Be careful to get the signs right in this *IV* column as some will be positive and others negative throughout.

After completing this operation, the vertical columns of the dates and the Initial Variations are transferred to a horizontal table. We then proceed thus:

(1) Total each vertical quarter column and average the result by dividing by however many numbers there were in the column. Take care! There may not be the same number of figures in each quarter column!

(2) Add together the answers. This result has now to be forced to come to zero so if it is positive we must subtract a little from each of the four and if it is negative we must add a little to each of the four. If it came to +2 we would take ½ off each of the four column totals. If it came to −4 we would add 1 to each of the four column totals and so on. Having made this adjustment we check that the four column totals now add up to zero and, if they do, then we have the seasonal variations.

EXAMPLE

Here is a set of dates and appropriate Initial Variations:

Year and Quarter			*Initial Variations*
1963	4		—
1964	1		—
	2		−2
	3		−3
	4	*Here will appear the*	+6
1965	1	*calculations leading*	−10
	2	*to the trend. Each*	−1
	3	*trend will then be*	−4
	4	*subtracted from its*	+7
1966	1	*original to give the*	−11
	2	*initial variations on*	+4
	3	*the right.*	−7
	4		+8
1967	1		−12
	2		+3
	3		−10
	4		+11
1968	1		—
	2		—

We now transfer this information to a horizontal table and proceed as outlined above and illustrated below.

Year	*Quarter 1*	*Quarter 2*	*Quarter 3*	*Quarter 4*
1964	——	−2	−3	+6
1965	−10	−1	−4	+7
1966	−11	+4	−7	+8
1967	−12	+3	−10	+11
Totals	−33	+4	−24	+32
Averaged	−11	+1	−6	+8

−11 +1 −6 +8 = −8 and as the total must be adjusted to come to zero it means we must, in this case, add 2 on to each of the seasons thus

Adjusted	−9	+3	−4	+10

This gives −9 +3 −4 +10 which does check to zero so those figures are the Seasonal Variations.

SEASONALLY ADJUSTED FIGURES

Now if we are further required to seasonally adjust certain of the original figures in the light of our calculations of the Variations we must remember that the −9 referring to each first quarter means that the figure will be 9 below the true figure if seasonal variations were cut out. We must therefore remember to change the sign when making such adjustments.

EXAMPLE

Let us suppose that we are asked to adjust the original figures for 1967 in the light of our calculations and that the original figures were 29, 26, 20 and 25 we would adjust thus:

Quarter	*Original Figures*	*Seasonal Variation*	*Adjusted Figures*
1	29	−9	38
2	26	+3	23
3	20	−4	24
4	25	+10	15

21

Regression

If a set of data is such that it will produce a number of points scattered all over a graph when plotted, we call the resultant diagram a Scattergraph and there is little to be gained from trying to join the random points into a curve. There are two things we can do in such a case. We can either draw in freehand the line which seems to go nearest to most of the points. This is called a line of best fit. Or we can calculate in some way a line which seems mathematically sound and which is independent of such individual judgment. This is called a *Regression Line*. From either we can attempt to estimate what the value of the second variable is likely to be for any given value of the first and vice versa.

The simplest method of drawing a regression line is known as the three point method and to produce same we proceed thus:

(1) List the two variables in horizontal columns. Total and average each column. The two averages thus obtained form the first of the three points which will be joined to form our regression line.

(2) See where this first calculated point would be likely to appear in the original list if it had been included. Mark the point.

(3) Total and average in each column the readings above the point marked. This is our second calculated point.

(4) Total and average in each column the readings below the point marked. This is our third calculated.

(5) The three points thus calculated can now be plotted on a graph (with or without the original points according to the instructions in the question) and can be joined to form the regression line.

EXAMPLE

'Ten boys in a class did two tests and the marks they received are listed below. Calculate the three points which you would use to plot a regression line from the marks given.'

	Student									
	1	2	3	4	5	6	7	8	9	10
Test A	2	3	5	8	9	10	12	13	17	17
Test B	3	14	7	3	11	15	8	7	10	13

In other types of question as well as those on Regression, the figures are sometimes given in horizontal columns. In most cases it is advisable to rewrite them vertically before beginning your calculations.

	Test A	*Test B*
	2	3
	3	14
	5	7
	8	3
	9	11
	→	
	10	15
	12	8
	13	7
	16	10
	17	13
TOTAL	95	91
AVERAGE	9·5	9·1=First calculated point

The arrow indicates the place where the first calculated point would have appeared so we now total and average all the points above the arrow:

TOTAL	27	38
AVERAGE	5·4	7·6=Second calculated point

Then we total and average all the points below the arrow:

TOTAL	68	53
AVERAGE	13·6	10·6=Third calculated point

So for our regression line, we plot 9·5 as against 9·1; then 5·4 against 7·6; then 13·6 against 10·6 and join the three into a straight line. In Figure 13, the original points are plotted with + signs and the three calculated points with × signs. This is a good idea for general use. From the line we can read any × value for a given y value and vice versa.

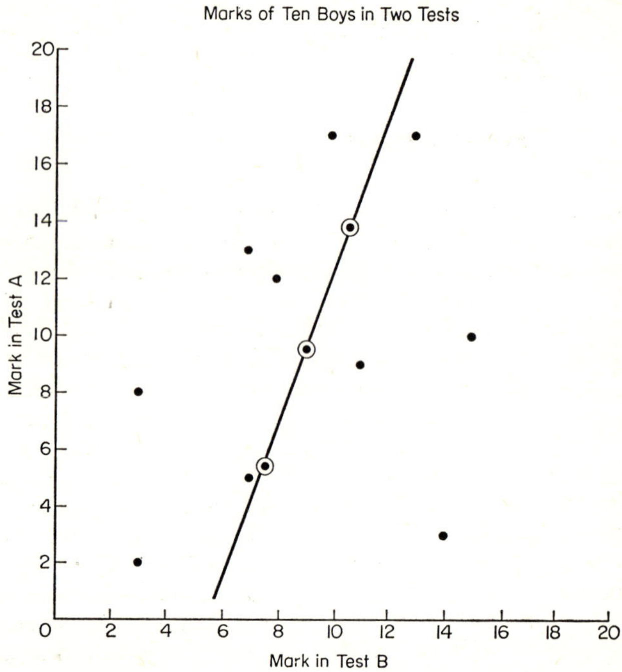

Fig. 13.—Regression by the 3 point method

22

Correlation

Clearly the accuracy of the regression lines depends upon the closeness of the relationship between the two variables. A means of measuring this closeness or *Correlation* exists and is called the *Coefficient of Correlation* (abbreviated to the small letter 'r'). The range of values which this coefficient can have runs from −1 through 0 to +1 and we are often asked to interpret our answers so we must know, first of all, what the different values mean.

−1 means that as the value of one variable increases, the value of the other shows a corresponding decrease. Lesser negative values show a diminishing tendency for the second value to decrease as the first increases. 0 or values slightly either side of zero means that there is no relationship at all. In other words if the first variable increases we do not know for sure what will happen to the other. The positive values indicate a growing tendency for the second variable to increase as the first increases until at +1 it increases correspondingly with increases of the first. −1 is known as perfect negative correlation and the scale then runs through high negative correlation, low negative correlation, zero correlation, low positive correlation, high positive correlation to +1 which is known as perfect positive correlation.

There are a number of different methods of calculating 'r' but we only need to deal with two of them. These are as follows:

COEFFICIENT OF RANK CORRELATION (r′)

Where the two variables are such that each can lead to the readings being placed in an order of merit we use the coefficient of rank correlation abbreviated to r with a little dash above it (r′). An example of this would be two sets of examination marks in two different subjects. In such a case each set of marks could be transformed into a class order as shown in the example which follows the method notes.

Method

(1) If the order of merit is not given in the question, work it out.

(2) Set out the two orders in ranks in two columns and add

a third column (d—for deviation) in which the two positions on each line are subtracted from one another regardless of sign.

(3) Square each figure in the deviation column to produce a fourth (deviation squared) column (d^2). Total this column.

(4) Calculate r′ using the formula

$$r' = 1 - \frac{\text{SIX TIMES THE SUM OF THE DEVIATION SQUARED COLUMN}}{\text{NUMBER OF ITEMS IN ORIGINAL RANKS} \times \left(\text{NUMBER OF ITEMS SQUARED MINUS ONE}\right)}$$

or

$$r' = 1 - \frac{6\Sigma d^2}{n(n^2-1)}$$

Now let us try this method out on the following example. 'A class of boys scored the following marks in two examinations:

Accounting	*Statistics*
60	80
70	65
75	75
65	70
55	60
50	50
45	55

Calculate the coefficient of correlation between the ranks and comment on the relationship present.'

Follow the method notes above in the worked out question below.

1/2/3	*Accounting Position*	*Statistics Position*	*d*	d^2
	4	1	3	9
	2	4	2	4
	1	2	1	1
	3	3	0	0
	5	5	0	0
	6	7	1	1
	7	6	1	1
				16 (Total)

4 Therefore $r' = 1 - \frac{6 \times 16}{7\ (49-1)} = 0{\cdot}71$

(because the sum of the d^2 column was 16 and there were seven sets of readings in our first two columns so n is 7 and n^2 is 49)

One would comment on such a result as follows: 'The answer shows a high positive correlation of 0·71 which means that as one variable increases the other will have a marked tendency to increase. There will thus be a good chance that a person who is low in the Accounting class will be low in Statistics whereas those who are good at Accounting will tend to be good at Statistics too.'

COEFFICIENT OF CORRELATION (r)

Where the two variables are such that it would be pointless to attempt an order of merit within each set, then we use the plain coefficient which can be calculated thus:

Method

(1) Rule seven columns and place the figures you are given in column one and four.

(2) Total the two columns individually and work out for each an average by dividing that total by however many numbers there are in the columns.

(3) In column 2 (we call this x) write down with sign the answer you get when you subtract the figure in the column from the column average. Do the same in column four for the second variable and call this column y.

(4) In column three, square each value of x and in column six, square each y value. Total these two columns—x^2 and y^2.

(5) In column seven multiply each x value by the corresponding y value thus producing an xy column. Total this column.

(6) Calculate 'r' using the formula:

$$r = \frac{\text{SUM OF THE XY COLUMN}}{\text{SQUARE ROOT OF}\left(\begin{array}{l}\text{SUM OF THE X SQUARED COLUMN MULTIPLIED BY THE}\\ \text{SUM OF THE Y SQUARED COLUMN}\end{array}\right)}$$

or

$$r = \frac{\Sigma xy}{\sqrt{(\Sigma x^2 \times \Sigma y^2)}}$$

EXAMPLE

'Two variables are given below. Work out the coefficient of correlation between them and comment upon the high negative correlation present.'

	Variable 1	x	x^2	*Variable 2*	y	y^2	xy
	1	−4	16	90	30	900	−120
	2	−3	9	80	20	400	−60
	3	−2	4	85	25	625	−50
	4	−1	1	75	15	225	−15
	5	0	0	55	−5	25	0
	6	1	1	45	−15	225	−15
	7	2	4	50	−10	100	−20
	8	3	9	40	−20	400	−60
	9	4	16	20	−40	1600	−160
TOTALS	45		60	540		4500	−500
AVERAGE	5			60			

From the formula therefore $r = \dfrac{-500}{\sqrt{(60 \times 4500)}} = -0{\cdot}96$

Here we have an example of high negative correlation which means that *as one variable rises, the other falls by an almost corresponding amount* so that we can expect variable two to decrease at about the same rate that variable one increases.

23

Hints and Tips

Examination questions are designed, not only to see that a candidate knows the method of doing some calculation, but also that he knows why we do it. Consequently there will often be such phrases as 'Discuss your results' or 'What do we mean by . . .'. It is therefore most important to read the whole question through very carefully. Many easy marks have been lost because a candidate has become so engrossed with a calculation and has been so delighted at obtaining a reasonable result that he has failed to notice some further instruction tucked away at the end of the question. Exercise 4C in the next chapter illustrates this principal very well. Now let us discuss some of the rather silly mistakes which crop up year after year in completed scripts:

(1) We draw a block diagram from a plain frequency and an ogive from a cumulative frequency. It is not usual to have an ogive from a plain frequency or to draw a block diagram from a cumulative frequency.

(2) When calculating the Median by the arithmetic method the easiest way of deciding what the group interval is will be to look how often a new group starts rather than to see the range of one group. If the first group is 6d. to 11d. and the second is 1/- to 1/5d., then the group interval is 6d. It is NOT 5d.

(3) Many students give the position of the median as the answer for the median itself. Remember that the position just tells us where the median lies and we have still to calculate it. This rule and rule 2 apply also to the 'Iles'.

(4) Remember to change the group interval at the end of each cycle in a ratio scale graph.

(5) In a weighted average index number the total of the Index × Weight sums is divided by the sum of the weights and not the number of items.

(6) When adjusting the totals of the quarter columns to zero in a seasonal variation calculation take care with the signs. Minus nine plus three will be minus six NOT minus twelve. Minus nine minus two will be minus eleven NOT minus seven.

The author has known years when these six simple errors have accounted for 50 per cent of the mistakes in a set of papers!

STANDARD DEVIATION FORMULA

Students who find mathematics very difficult are usually capable of transposing the various figures into the formula format but do not know how to work it out. In such a case the following steps are suggested. The formula is repeated first as a reminder.

$$\text{GROUP INTERVAL}\sqrt{\frac{\text{SUM OF THE } fd^2 \text{ COLUMN}}{\text{SUM OF THE } f \text{ COLUMN}} - \left(\frac{\text{SUM OF THE } fd \text{ COLUMN}}{\text{SUM OF THE } f \text{ COLUMN}}\right)^2}$$

(1) Divide the sum of the fd column by the sum of the f column.

(2) Square the result.

(3) Divide the sum of the fd^2 column by the sum of the f column.

(4) Subtract the answer you obtained in (2) from that obtained in (3).

(5) Find the square root of the result.

(6) Multiply that answer by the group interval.

OPEN END GROUPS

If the groups in a frequency distribution have an open end—that is to say the last one has no definite end, an end should be estimated from the other groups for calculation purposes. Thus in the following example:

11-20	it would be reasonable to suppose that
21-30	
31-40	51-60 be taken as the final group for
41-50	
51 and over	calculation purposes if necessary.

UNEQUAL GROUPS

If the range of the groups is unequal some similar intelligent supposition should be made to even them. This will sometimes involve an intelligent split of the frequency also.

Groups	f
0-20	17
21-30	11
31-40	13
41-50	9

Here it would be reasonable to split the first group into two—0-10 and 11 to 20. The frequency will then have to be split reasonably. To follow the normal distribution the figures would be rising so a split into 8 and 9 or 7 and 10 would be reasonable. Any *reasonable* method will gain full marks.

GENERAL NEATNESS

There is not usually any specific proportion of marks for neatness but the Statistician is always expected to produce copperplate work and your examination script must reflect the way in which you would present figures to your boss at work.

All diagrams should be fully labelled. Answers and conclusions must be obvious to the reader. Lines on graphs should neither be tramlines nor invisible. Remember that plus signs look neater for marking points on graphs than X signs as the plus signs lie over rules lines on the paper.

Take care to complete the information requested on the front of the examination answer book and, in particular, see that the questions you have attempted are marked in their correct order in the box which often appears for the purpose of assessment on the cover.

24

Exercises (Fourth Set)

The past examination questions in this chapter cover the work done in Chapters 19 to 23.

4A What is meant by the Trend? From the following data select a suitable moving average to find the trend. On the same graph draw the curves of the trend and of the original data.

ELECTRICITY GENERATED IN THE U.K.
IN 000 MILLION KWH.

Year	*Quarter 1*	*Quarter 2*	*Quarter 3*	*Quarter 4*
1960	116·4	116·2	123·1	126·3
1961	123·5	128·1	128·8	137·3
1962	143·2	141·3	140·4	149·6
1963	165·3	149·1	150·0	156·9

4B From the data below compute the trend stating your method of determining the cycle. Plot it with the original series on a graph. Explain how the procedure for calculating a moving average differs in cases where the period contains an even number of years to when it contains an odd number of years.

Year		*Year*		*Year*		*Year*		*Year*	
1940	22·2	1945	18·4	1950	18·8	1955	28·4	1960	25·8
1941	26·2	1946	22·8	1951	20·0	1956	25·2	1961	23·9
1942	16·6	1947	23·0	1952	26·0	1957	20·2	1962	29·0
1943	20·0	1948	24·6	1953	22·8	1958	24·8	1963	22·1
1944	20·2	1949	22·0	1954	24·6	1959	21·2	1964	23·2

4C BANKERS CLEARING HOUSE RETURNS: VALUE OF CREDIT CLEARING IN £M

Year	*1st Quarter*	*2nd Quarter*	*3rd Quarter*	*4th Quarter*
1961	1193	1308	1321	1430
1962	1410	1512	1474	1625
1963	1592	1730	1726	1886
1964	1864	1984	1959	2081

Calculate the trend of credit clearing using a suitable moving average. Hence estimate the seasonal variation in the series and comment on your result.

4D The quarterly index numbers of wholesale prices for the four years 1957 to 1960 are tabulated below. Calculate the trend and the seasonal variation using the moving average method. Eliminate the seasonal variation from the 1959 figures and explain why your deseasonalized values differ from the trend values at the corresponding time.

INDEX NUMBERS OF WHOLESALE PRICES (AVERAGE 1866 = 100)

Quarter	*1957*	*1958*	*1959*	*1960*
1	393	356	354	354
2	382	363	353	359
3	365	350	354	348
4	364	360	357	360

4E Ten pupils sat two Mathematics tests. The marks in the two tests were as follows:

Pupil	*A*	*B*	*C*	*D*	*E*	*F*	*G*	*H*	*I*	*J*
Test 1	15	17	15	6	8	14	7	6	3	12
Test 2	14	17	14	8	5	14	5	8	4	10

Calculate the coefficient of rank correlation between the two sets of marks.

(A.E.B.)

4F Three judges in a beauty contest placed the contestants in the following orders of merit:

Contestant:	*A*	*B*	*C*	*D*	*E*	*F*	*G*	*H*
Judge 1	4	3	1	6	7	8	2	5
Judge 2	6	4	5	7	2	1	3	8
Judge 3	6	7	5	8	4	3	1	2

Calculate the rank correlation coefficients between the three pairs of judges. If you were the referee which of the contestants would you place first, second and third?

(A.E.B.)

4G What is the difference between correlation and rank correlation?

Within what range does the coefficient of correlation usually lie?

What can we say about a reading of 0?

What can we say about readings at both extremes of the range?

The table below gives the marks in Statistics and Accounts of seven boys on an O.N.D. course. Place them in order and calculate the coefficient of rank correlation.

Boy	*Statistics*	*Accounts*
ABRAHAMS	57	88
BARLOW	66	53
CHANCELLOR	79	84
DRONFIELD	55	77
EKERSLEY	40	50
FIELDHOUSE	46	41
GOLDBERG	37	43

4H The following table shows the result of an experiment in which readings of x were compared with readings of y. Plot the values on a scatter diagram and fit a regression line by the three point method showing all working.

X	75	80	85	90	95	100	105	110	115	120
Y	554	550	548	544	541	536	533	529	526	522

Explain what is meant by the following terms:

(*a*) Line of best fit.
(*b*) Regression line of X on Y.
(*c*) Regression line of Y on X.

4J An experiment was conducted to find the relation between two quantities X and Y. The results obtained were as follows:

X	2	3	4	5	10	15	20
Y	1·4	1·2	1·1	0·8	0·8	0·6	0·7

Change to a new pair of variables X ft. and Y ft. where X ft. = $\frac{1}{X}$ and Y ft. = Y. Plot a scatter diagram for Y ft. against X ft. and fit a regression line by the three point method.

(A.E.B.)

4K A general intelligence test was given to fifteen boys of different ages and the results are shown in the table below:

Boy	*Age in Months*	*Marks*
a	134	62
b	141	63
c	154	64
d	144	62
e	163	72
f	164	69
g	175	80
h	185	76
i	175	72
j	188	82
k	185	80
l	202	82
m	205	82
n	202	85
o	200	81

Plot a scatter diagram and draw a regression line by the three point method. State with reasons which boy could be considered the most intelligent and the least intelligent taking age into consideration.

(A.E.B.)

25

Ten-minute Tests

The questions in these tests are intended to provide practice in the operation of the various formulas and techniques discussed in this book. They are not meant to resemble examination questions. Students will probably need more than ten minutes for each test initially but should aim to complete each in ten minutes before sitting their examination.

TEST 1

(1) In a certain distribution the group interval was 2, the group of zero deviation was 9 to 11, the sum of the frequency times deviation column was 100 and the number of items in the distribution was 100. Calculate the arithmetic Mean.

(2) Add to the above the fact that the sum of the frequency times deviation squared column was 500 and work out the Standard Deviation.

(3) Two groups in a distribution were '8d. to 10d.' and '11d. to 1/1d.'. Their cumulative frequencies were 35 and 55 respectively and the position of the Median was 50. Work out the Median.

(4) If a Lower Quartile was 7d. and the Upper Quartile was 1/1d. what would be the Quartile Deviation?

(5) Commodity Q had an Index of 100 in 1955 and cost 10d. If it cost $11\frac{1}{2}$d. in 1962 what would have been the Index Number then?

(6) A book cost £1 in 1950 but now costs 25/-. What is the Index Number for the book now based on 1950?

(7) What would the Index Number for the same book be for 1950 based on today?

(8) Commodity S used to cost 6d. and now costs $10\frac{1}{2}$d. If the first price was taken in the base year and the Index times Weight column for that commodity came to 3500, what weight was used for S?

(9) What readings in whole numbers only would you expect to be covered by a group labelled '40 to the nearest ten'?

(10) In a rank correlation the sum of the deviation squared column was 110 and the number of items in the original ranks was 10. Calculate the coefficient of rank correlation.

TEST 2

(1) What actual readings in whole numbers only would you expect to be covered by a group entitled '40 to the nearest ten below'?

(2) In a certain distribution the group interval was 10, the centre of the nought deviation group was 30, the sum of the frequency times deviation column was 150, the sum of the frequency times deviation squared column was 1250 and the number of items was 50. What was the arithmetic Mean?

(3) From the above figures calculate the Standard Deviation.

(4) The totals of four initial variation columns were −100, 100, −220 and 140. If each applied to four readings, calculate the seasonal variations.

(5) If the first two figures on the first cycle of a semi logarithmic graph's vertical scale were 1 and 2, what would the first two figures of the fifth cycle be?

(6) A commodity cost 3/- in 1949 and 4/- in 1959. Today the Index based on 1959 was 125, what would the Index for today based on 1949 be?

(7) What actual readings in whole numbers only would you expect to be covered by a group labelled '90 to the nearest ten above'?

(8) Give each of the following numbers to two significant figures. 2709, 361, 279, 37503.

TEST 3

(1) If four Initial Variation columns totalled 140, −60, 200 and −150 and the first three referred to four originals and the last to three originals, work out the seasonal variations.

(2) If a machine costs £300 now and cost £250 in 1959, what is the Index for today based on 1959?

(3) If the first two figures of the third cycle on the vertical scale of a semi-logarithmic graph were 10000 and 20000, what must the first two figures of the first cycle have been?

(4) Two groups in the middle of a cumulative distribution were £31 and £40 and £41 to £50 with cumulative frequencies of 350 and 450 respectively. If the position of the Median was 360, calculate the Median.

(5) In a distribution the centre of the zero deviation group was 9, the group interval was 7, the number of items was 100, the sum of the frequency times deviation column was 300 and the sum of the frequency times deviation squared column was 5800. Calculate the arithmetic Mean.

(6) Calculate from the above figures the Standard Deviation.

(7) If a commodity cost 16/8d. in 1959 and now costs 20/10d., give an Index for today based on 1959.

(8) Calculate to the nearest whole number an Index for 1959 based on today.

TEST 4

(1) In a distribution the centre of the nought deviation group was 15, the sum of the frequency times deviation column was 200, the sum of the frequency times deviation squared column was 2000, the group interval was 10 and the number of items 40. What was the arithmetic Mean?

(2) Calculate the Standard Deviation from the above.

(3) If the first cycle of a semi-logarithmic graph begins at 1, what does the 7th cycle begin at?

(4) An item cost 4d. in 1950 and now costs 6d. Give an Index for today based on 1950.

(5) Give an Index for 1950 based on today.

(6) A group from 0 to 10 had a cumulative frequency of 16 and the next group from 11 to 20 had a cumulative frequency of 37. If the position of the Median was $26\frac{1}{2}$, what was the Median.

(7) What whole number readings would you put in a group labelled '50 to the nearest ten above'?

(8) In a distribution the centre of the 0 deviation group was 25, sum of the frequency times deviation column was 10, the sum of the frequency times deviation squared column was 20, the number of items was 5 and the group interval was 5. Find the Mean.

(9) Calculate the Standard Deviation from the above.

(10) Hence find the Coefficient of Variation.

TEST 5

(1) If the sums of four initial variation columns are −12, 8, 24 and −36, each referring to four original figures, what are the seasonal variations?

(2) If the totals were −32, 56, −4 and 20 and each referred to four originals, what would the seasonal variations be?

(3) Calculate the limits involved by '350 to the nearest 50'.

(4) Calculate the limits involved by '70 to the nearest ten above'.

(5) Calculate the limits involved by '4·49 to two places of decimals'.

(6) Calculate the limits involved by '2000 to two significant figures'.

(7) If a distribution has a centre of zero deviation at 15, sum of

frequency times deviation column at 100, sum of frequency times deviation squared column at 400, group interval at 10 and fifty items in the distribution, what is the Mean?

(8) From the above calculate the standard deviation.

TEST 6

(1) In a cumulative distribution a group 16 to 20 had a cumulative frequency of 300 and the group from 21 to 25 had a cumulative frequency of 350 with a Median position of 330. What was the Median?

(2) If an Upper Quartile was 33 and the Lower Quartile was 7, what was the Quartile Deviation?

(3) If four Initial Variation columns of four numbers each totalled 9, −3, 6 and −4, work out the seasonal variations.

(4) Give the limits of 51·756 to three places of decimals.

(5) In a distribution the centre of the o deviation group was 10, the sum of the frequency times deviation column was 300, the frequency times deviation squared column totalled 1300, the group interval was 10 and the number of times was 100. Calculate the arithmetic Mean.

(6) Calculate the Standard Deviation from the above.

(7) If an article cost 25/- in 1960 and now costs 26/3, give an Index for today based on 1960.

(8) If an earlier Index based on 1940 had reached 200 when replaced by the above in 1960, what would the old Index have reached today if it had remained in force.

Similar tests can easily be produced by following the pattern of the above examples. It is a good idea for students to try to make up one for themselves so that all the calculations can be done mentally and they can try out their tests on other members of the class.

Index